LIGHTING

SIXTH EDITION

D. C. PRITCHARD

 LONGMAN

Addison Wesley Longman Limited
Edinburgh Gate, Harlow
Essex CM20 2JE, England
and Associated Companies throughout the world

First published 1969
Second edition 1978
Third edition 1985
Fourth edition 1990
Fifth edition 1995
Sixth edition 1999

British Library Cataloguing in Publication Data
A catalogue entry for this title is available from the British Library

ISBN 0–582–35799–3

Set by 35 in 10/12pt Ehrhardt
Produced by Addison Wesley Longman Singapore (Pte) Ltd.
Printed in Singapore

CONTENTS

Acknowledgements   vi

1 THE LANGUAGE OF LIGHT   1
2 UNITS OF LIGHT   17
3 COLOUR   49
4 LAMPS   63
5 LUMINAIRES   99
6 DAYLIGHT   123
7 DESIGN OF GENERAL LIGHTING SCHEMES   147
8 LIGHTING OF SPECIFIC BUILDING TYPES   187
9 ENERGY MANAGEMENT AND LIGHTING   197
10 ROADWAY LIGHTING   211
11 FLOODLIGHTING   235

Appendix A   Computer programs   245
Appendix B   Legal requirements   250
Appendix C   Range of typical lamps and their lumen outputs   253

Bibliography   255
Answers to self-assessment tasks   259
Numerical solutions to exercises   261
Index   263

ACKNOWLEDGEMENTS

The author is grateful to the following for permission to reproduce copyright material:

Philips Lighting Ltd for Figs 2.6, 7.7 and 7.8 and Tables 11.1 and 11.2; Sylvania Lighting Products Ltd for Fig. 4.5; Thorn Lighting Ltd for Figs 5.13, 7.9, 7.10, 8.2–8.8, 9.10 and 10.14 and Tables 7.3 and 7.10; Building Research Establishment for Fig. 6.2 and Tables 6.1–6.8, Crown copyright; Pilkington Glass Ltd for Table 6.10; The Electricity Association for Figs 9.5 and 9.6; Whitecroft Lighting Ltd for Fig. 9.7; The Hyper Light Partnership for Fig. A2; Thorn Lighting Limited for Plate 5; The Highland Council for Plate 6; VIEW for Plate 7; Windsor and Maidenhead Council for Plate 8.

The Chartered Institute of Building Services Engineers for Figs 1.6 and 7.5 and Tables 6.9, 7.1, 7.2, 7.8 and 8.1–8.4 taken or adapted from the *CIBSE Interior Lighting Codes* and the *CIBSE Window Design Guide*.

Extracts from British Standards are reproduced with the permission of BSI. Complete copies can be obtained by post from BSI Customer Services, 389 Chiswick High Road, London W4 4AL; tel. 0181 996 7000.

The cover photograph is reproduced courtesy of Paul Bock.

The guidance notes on light pollution are from the publication of the Institution of Lighting Engineers.

The author would like to thank John Pickup and Dr M. B. Kostic of the University of Belgrade for the very constructive comments they made on the previous edition.

CHAPTER 1

THE LANGUAGE OF LIGHT

TOPICS COVERED

ELECTROMAGNETIC RADIATION
Electromagnetic spectrum

VISUAL RESPONSE
Relative response of the eye
Converting from radiation to light units

VISUAL PERFORMANCE OF THE EYE
Optical performance of the eye

COLOUR PERCEPTION
General condition of the eye
Relative brightness – contrast sensitivity
Glare from the task and its surrounds
Movement in the task

VISUAL TASK REQUIREMENTS

VISUAL PREFERENCE

People's attitudes to lighting design are paradoxical. Those people who have never considered the subject in any depth accept it as part of life and living, requiring little more thought than the strategic placing of some 60 W 'bulbs' and the operation of a switch. And, to a point, they are right. After all, there are very few situations where there is insufficient light by which to see. The danger occurs when a person with this attitude has to make responsible lighting decisions on behalf of other people. The paradox is that as soon as a person becomes involved in lighting he or she enters a new science which demands imagination and engineering ability. To be successful, the person must be creative in his or her ability to reveal and display the visual scene. The person must also have confidence in the calculations and an understanding of the available lighting tools. It is perhaps little wonder that electric lighting seldom compares favourably with daylight.

The first question could well be: 'What is light?' Strictly speaking, it is purely a human sensation in similar fashion to sound, taste, smell and warmth. Something is necessary to stimulate the senses, and in this case it is electromagnetic radiation falling on the retina of the eye. Light can therefore be considered as a combination of radiation and our response to it.

ELECTROMAGNETIC RADIATION

The basic atomic structure is a positively charged nucleus around which orbits an array of negatively charged electrons, as shown in Fig. 1.1. If the atom receives energy by collision with another atom or by heat, it will release this energy via its

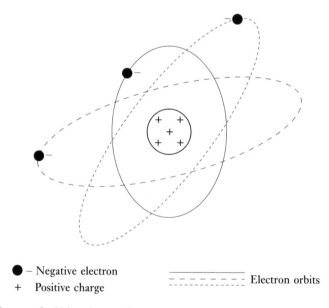

● – Negative electron
+   Positive charge

– – – – – – –   Electron orbits
- - - - - - - - -

Fig. 1.1   Pattern of orbiting electrons in an atom

electrons – hence *electro*magnetic. This radiation obeys various 'laws', travels in straight lines and consists of individual packets of energy called *photons*. The energy of a photon is

$$eV = h\nu \qquad\qquad [1.1]$$

where h is Planck's constant, ν is the frequency of the radiation, and eV is the energy in electronvolts, i.e. the work required to move an electron through a potential difference of 1 volt.

Example

A mercury electron emits photons whose energy level is 4.89 eV. If Planck's constant is 6.62×10^{-34} J s calculate the wavelength of the emitted radiation.

Solution

The frequency of the radiation in *hertz* (cycles per second), abbreviated Hz, can be found from the formula

$$\nu = \frac{eV}{h}$$

When substituting values, the units must be correct. If h is in joule seconds, then eV must be in joules. Thus

$$1\ eV = 1.603 \times 10^{-19}\ J$$

$$\therefore \quad \nu = \frac{4.89 \times (1.603 \times 10^{-19})}{6.62 \times 10^{-34}}$$

$$= 1.18 \times 10^{15}\ Hz$$

To convert to wavelength, for all electromagnetic radiation

$$\nu = \frac{c}{\lambda} \qquad\qquad [1.2]$$

where c is the velocity of the photon in a vacuum (approx. 3×10^8 m/s) and λ is the wavelength in metres. Hence

$$\lambda = \frac{c}{\nu}$$

$$= \frac{3 \times 10^8}{1.18 \times 10^{15}}\ m$$

$$= 254 \times 10^{-9}\ m$$

$$= 254\ nm$$

This is the wavelength of the strong short-wave ultraviolet radiation from mercury discharge.

Self-assessment task 1.1

If the principal radiation of a low-pressure sodium lamp is at a wavelength of 589.6 nm, calculate the frequency of its radiation.

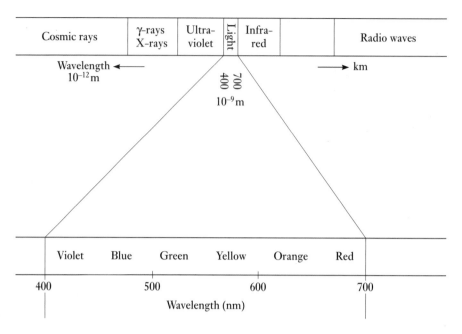

Fig. 1.2   The electromagnetic spectrum

Electromagnetic spectrum

Electrons are the source of radiation of varying frequencies and wavelengths and this radiation is used for many purposes other than light. It helps to decide whether to think in terms of wavelength or frequency.

As already seen in eqn [1.2],

$$\text{Wavelength} \propto \frac{1}{\text{frequency}}$$

In this book wavelength will be used to avoid confusion. The spectrum or range of wavelengths is shown in Fig. 1.2. The names of the various bands of the spectrum indicate, in general, how the energy is used. Light occupies a very narrow band of this spectrum (between 400 and 700 nm) and that band can be further subdivided into the range of colour sensations that most people experience.

One interesting aspect is the energy of a photon. This was given in eqn [1.1] and can be rewritten as

$$\text{Energy} \propto \frac{1}{\text{wavelength}} \tag{1.3}$$

Hence, as the wavelength gets shorter the energy of the photon increases. If this is related to the effect of radiation on people, the results can be quite devastating at wavelengths shorter than 300 nm. At longer wavelengths, in the millimetre and metre bands, the effect is negligible until it is transduced by a radio or television receiver.

VISUAL RESPONSE

The eye is a magical part of the body. It conveys the bulk of our information and governs the most precious of our five senses.

In the process of seeing it performs two quite separate functions. First, it acts as a sophisticated variable lens system focusing a clear pattern of the scene on the back of the eye – the retina. The components of the eye are shown in Fig. 1.3. At this stage it is as well to remember that all wavelengths of energy are accepted and eyes must be protected in some way against harmful radiation which is invisible, i.e. lies outside the 400–700 nm band.

Secondly, that pattern is converted via the nerve endings in the retina and through the optic nerve as a complex electrical signal to the brain. The process of conversion is not fully understood, but is referred to as a photochemical change where the photons of energy cause a chemical change in the retina.

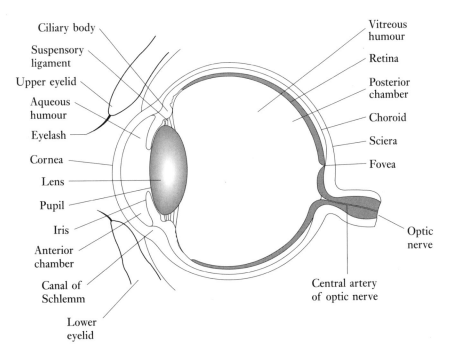

Fig. 1.3 Sectional diagram of the human eye

Table 1.1   The luminous efficiency functions for photopic vision (V_λ)

λ (nm)	V_λ	λ (nm)	V_λ
400	0.000	550	0.994
410	0.001	560	0.995
420	0.004	570	0.952
430	0.011	580	0.870
440	0.023	590	0.757
450	0.038	600	0.631
460	0.060	610	0.503
470	0.090	620	0.381
480	0.139	630	0.265
490	0.208	640	0.175
500	0.323	650	0.107
510	0.503	660	0.061
520	0.710	670	0.032
530	0.862	680	0.017
540	0.954	690	0.008
		700	0.004

Source: Principles of light measurements, CIE Publication No. 18, 1970.

Relative response of the eye

The nerve endings in the retina are of two types, cones (photopic vision) and rods (scotopic vision). The latter are highly sensitive and only effective at very low lighting levels, well below those to be discussed in this book, and will not be considered further.

The cones respond over a waveband from 400 to 700 nm and this response varies in both quantity and quality. Table 1.1 gives the relative response in terms of quantity, and the qualitative response in terms of colour has been shown in Fig. 1.2.

Table 1.1 shows (as one would expect) that there is a gradual increase to maximum response at 555 nm and then a gradual decrease. Plotting λ against V_λ will illustrate this and provide a useful reference curve.

Converting from radiation to light units

The radiation entering the eye is measured in watts. It then undergoes transformation because some wavelengths are more effective than others. The unit of light power is the *lumen* (abbreviated lm) and the general relation between the two units is

$$\text{Light power} = K\int_{\lambda=\infty}^{\lambda=0} P_\lambda V_\lambda \text{ lm } d\lambda \tag{1.4}$$

where K is a factor normally accepted as 675, P_λ is the radiation power at wavelength λ and V_λ is the luminous efficiency of the eye at wavelength λ (see Table 1.1).

Example

If a light source emits 60 W at 589 nm, how many lumens does it emit?

Solution

At 589 nm, P_λ is 60 W (given in the problem) and V_λ is 0.768 (from Table 1.1). Therefore,

$$\text{Light power} = 675 \times 60 \times 0.768 \text{ lm}$$
$$= 31\ 104 \text{ lm}$$

This is a simple calculation when considering a single wavelength and requires more thought when the source is not a single wavelength (monochromatic) but a range of wavelengths (heterochromatic)

Example

Calculate approximately the lumens emitted from the following table of radiant power distribution for a 200 W filament lamp.

Waveband (nm)	<400	400–450
Power (W)	2	1
Waveband (nm)	450–500	500–550
Power (W)	1.5	2
Waveband (nm)	550–600	600–650
Power (W)	2.5	3
Waveband (nm)	650–700	>700
Power (W)	3.5	170

Solution

This can be done in tabular form. The degree of accuracy will depend on the number of bands taken. In this case 50 nm bands are really too wide but an approximate answer is asked for.

Waveband (nm)	P_λ (W)	Mid value V_λ	$P_\lambda \times V_\lambda$
<400	2	0	0
400–450	1	0.008	0.008
450–500	1.5	0.110	0.160
500–550	2.0	0.780	1.560
550–600	2.5	0.910	2.270
600–650	3.0	0.320	0.960
650–700	3.5	0.020	0.070
>700	170	0	0
		Total =	5.03

$$\therefore \quad \text{Light power} = 675 \times 5.03 \text{ lm}$$
$$= 3395 \text{ lm}$$

The efficiency of the lamp is expressed as *luminous efficacy* and is the ratio of lumens to lamp watts. In this case,

$$\text{Luminous efficacy} = \frac{3395}{200}$$

$$= 17.0 \text{ lm/W}$$

It is worth noting that although a filament lamp is an efficient radiator, the large proportion of this radiation is beyond 700 nm in the infrared band and hence useless. Filament lamps are basically inefficient.

Self-assessment task 1.2

A lamp emits *all* its radiant energy equally between 400 and 700 nm. The remaining energy (50 per cent) is conducted and convected. Calculate the approximate luminous efficacy of the lamp.

VISUAL PERFORMANCE OF THE EYE

A number of factors affect our ability to see and these can be subdivided into the human factors and the environmental factors. Human factors include:

- optical performance of the eye
- colour perception
- general condition of the eye

Environmental factors include:

- relative brightness of the task and its surrounds
- glare from the task or surrounds
- movement in the task

Optical performance of the eye

When the lens system is normal the eye should recognise detail of angular size 0.5 minutes of arc or less. 'Angular' size is the 'size' as seen and is a function of both size and viewing distance. This is illustrated in Fig. 1.4 using a standard

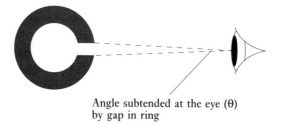

Angle subtended at the eye (θ)
by gap in ring

Fig. 1.4   The Landolt broken circle

test symbol called a 'Landolt broken circle'. An optician will test a person's sight by the ability to recognise letters on a standard test chart. A person who has 6:6 vision can recognise an 8.75 mm letter at a distance of 6 m; and 6:12 means the person can only read at 6 m what a 6:6 can read at 12 m, i.e. the person's eyesight is substandard.

In experiments on vision and visual performance the process of recognition is normally described in terms of *visual acuity* where

$$\text{Visual acuity} = \frac{1}{\text{minimal angular size (in minutes)}} \qquad [1.5]$$

Therefore, the starting point in understanding lighting is to appreciate that the eye must be functioning properly.

Example

A signwriter has to erect a nameboard with the letters BLC easily discernible from a distance of at least 1 km. Allowing for average acuity of 2 and adding a safety factor of 2, determine the minimum height of the letters if the gap in the letter C is one–fifth the height of the letter.

Solution
For a visual acuity of 2

$$2 = \frac{1}{\theta} \qquad [*]$$

where θ is the angle subtended by the gap at the eye.

Let the height of the gap be x metres. If the angle θ is in radians, then

$$\theta = \frac{x}{\text{distance}} = \frac{x}{1000} \text{ rad}$$

but θ is in minutes and π radians is the same as 180×60 min. From [*] above, $\theta = \frac{1}{2}$ min; thus, converting minutes to radians,

$$\theta = \frac{1}{2} \times \frac{\pi}{180 \times 60} \text{ rad}$$

$$= 0.000145 \text{ rad}$$

$$\therefore \quad \frac{x}{1000} = 0.000145 \text{ rad}$$

and

$$x = 0.145 \text{ m}$$

Increasing by the factor of 2,

$$x = 0.29 \text{ m}$$
$$\therefore \quad \text{Letter height} = 5x$$
$$= 1.45 \text{ m}$$

COLOUR PERCEPTION

There have been many theories on colour vision, the trichromatic theory being the one most widely accepted today. In the chapter dealing with colour, the process of additive mixing of primary colours is discussed and it is shown that this can produce a complete spectrum of colours. Applying this to vision, the theory considers the cones to be grouped into three types, each type sensitive to different wavelengths of energy. The interaction of these three groups is then responsible for the stimulus which is interpreted by the brain as 'colour'. We may not all see colours in the same way although we will all describe energy at, for example, wavelength 650 nm as creating a 'red' sensation. This particular consideration may help to explain why people can have such varying tastes in colour.

Despite the fact that, under normal day vision, colour is always present, the units and design techniques in lighting calculations are not able to take colour fully into account. This is not an implication that colour is unimportant.

An understanding of colour is essential when discussing visual performance, for not only are some colours more visible than others, but the contrast between colours is important. Blue contrasts strongly with yellow as these colours are 'complementary', but not as strongly with green as these colours are close in the spectrum.

General condition of the eye

The eye deteriorates with age. Anyone born with normal eyesight can expect to be wearing glasses by the time they reach 40. This is mainly due to weakening of the lens muscles and a difficulty in being able to focus at short distances. This can be partially corrected by wearing glasses. The condition is known as *presbyopia*. More serious problems include cataract, which is the clouding over of the lens. This impairs clear vision and scatters the light entering the eye making the sufferer far more sensitive to glare.

It is difficult to take these kinds of defects into account in general lighting schemes, but where it is known that the average age will be high – e.g. an old people's home – then extra consideration must be given to the quantity of light and avoidance of glare.

Relative brightness–contrast sensitivity

An object is seen because of the difference in brightness between the object and its immediate background. For example, a white letter would be very clearly seen against a black background but would be virtually invisible against a white background. This is an obvious statement, but the visual process is not so obvious as to suggest that we always strive for maximum contrast. People who wear tinted sun-glasses do not think so, as they deliberately reduce the brightness of all they can see.

There are two main factors to consider:

1. The contrast between the task and its immediate surrounds, e.g. the letters and pages of this book.
2. The brightness of the general background, e.g. the table on which this book lies or what is seen around the book when reading.

When discussing mathematical relationships, objective terms are normally used. In this instance *brightness* is subjective and describes the appearance, *luminance* is the physical measurement of brightness. The unit of luminance is the candela per square metre (abbreviated cd/m^2):

$$\text{Contrast} = \frac{L_T - L_B}{L_B} \qquad\qquad [1.6]$$

where L_T is the luminance of the task and L_B is the luminance of the immediate background.

Example

Dark wool is viewed against a dark background. The illumination is the same on both wool and background. If the reflectance of the background is 3 per cent and that of the wool is 5 per cent, what is the contrast?

Solution

If the illuminance is constant, then the luminance is proportional to the reflectance, and from eqn [1.6]

$$\text{Contrast} = \frac{0.5 - 0.3}{0.3}$$
$$= 0.66$$

Contrast sensitivity is more difficult to explain. It refers to the ability to recognise contrast in the task and is basic to work on contrast rendition factors (CRFs). Considering the threshold condition – i.e. when the contrast is just visible – the difference in luminance can be expressed as ΔL. This again can be expressed as a ratio to the average luminance. Figure 1.5 shows an experimental relation between this threshold contrast and the general luminance of the field of view. It indicates an increase in sensitivity as the general luminance increases. What it does not show is that further increases in luminance will tend to decrease the contrast sensitivity due to the effects of glare.

Example

The general lighting level is increased to raise the general luminance level from 10 to 100 cd/m^2. What is the change in contrast sensitivity expressed in terms of contrast?

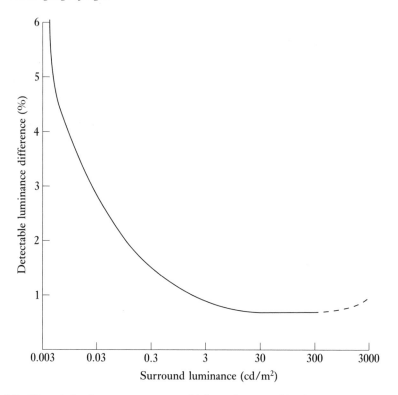

Fig. 1.5   The relation between contrast sensitivity and surround luminance

Solution

Using eqn [1.6] and considering the threshold condition, it can be deduced from Fig. 1.5 that

- at 10 cd/m² the contrast is approximately 0.75 per cent – condition 1;
- at 100 cd/m² the contrast is approximately 1.0 per cent – condition 2.

This indicates the minimum contrast that can be seen under specified lighting conditions. Applying eqn [1.6] to condition 1,

$$L_B = 10 \text{ cd/m}^2$$

The minimum perceptible contrast difference $\Delta(L_T - L_B)$ is given by

$$\frac{\Delta(L_T - L_B)}{L_B} = \frac{0.75}{100}$$

$$\therefore \quad \Delta(L_T - L_B) = 10 \times \frac{0.75}{100}$$

$$= 0.075 \text{ cd/m}^2$$

Applying eqn [1.6] to condition 2, .

$$L_B = 100 \text{ cd/m}^2$$

$$\frac{\Delta(L_T - L_B)}{L_B} = \frac{1.0}{100}$$

$$\therefore \quad \Delta(L_T - L_B) = \frac{100 \times 1.0}{100}$$

$$= 1.0 \text{ cd/m}^2$$

This shows that, as the lighting level is increased, the minimum perceptible luminance difference is increased. This is of little significance if the lighting is increased evenly. If the background luminance is increased, but not the task – e.g. by working on a white surface rather than a grey one – then this increase from 0.75 to 1.0 cd/m^2 represents a loss in visual performance. It is these aspects of visual performance that concern research workers such as H. C. Weston, H. R. Blackwell, P. Boyce and others. Specifying good lighting conditions is more than specifying the quantity of light. It involves the 'revealing power' in terms of the direction of light, surface properties of the task and general lighting of the environment.

Glare from the task and its surrounds

Glare is a development of contrast. Normally the eye adapts to whatever it is viewing, but if the task or background is too bright or the contrast is too great, vision suffers either by the situation becoming visually uncomfortable (*discomfort glare*) or by the task becoming difficult to see (*disability glare*), or both. Examples of glare unfortunately surround us, and can be far more disturbing than lack of illumination.

Common forms of discomfort glare are interior and exterior electric lighting installations. Chapter 7 explains how interior lighting schemes can be designed to keep this type of glare under control.

Disability glare can come direct from a light source or window, or by reflection off a glossy surface (*veiling glare*). The modern office with its array of visual display units is a common situation where this problem must be considered. CRF, referred to above, is one aspect of design where veiling glare is recognised and accounted for in the design process.

Self-assessment task 1.3

Describe the *main* type of glare experienced from the following:

- a motor car headlight
- a dirty windscreen of a car
- main road lighting

Movement in the task

Seeing is not a static process. The eye continually scans the field of view. We blink approximately 15 times per minute and the retinal response automatically adapts to the general lighting level. The lens shape changes to alter the focus and the iris expands and contracts to control the amount of light entering the eye.

The source of light often varies in quantity and quality, whether daylight or electric light. If electric, it will probably be flashing 100 times per second. So even in what appears to be a passive situation, such as sitting at a desk reading, there is a considerable element of movement and change.

Normal assessment of lighting requirements assumes that the task is stationary and there is sufficient time to focus on it. When the task is moving, such as work on a conveyor belt assembly, then extra lighting should be provided.

VISUAL TASK REQUIREMENTS

The two main aims of good lighting are:

1. To reveal the task effectively.
2. To reveal appropriately the general surround.

To achieve the first aim involves consideration of the factors already discussed, and Fig. 1.6 (taken from the CIBSE 1984 Lighting Code) shows how these factors can be related. This particular diagram relates visual performance to illuminance levels for three specific contrasts (C) and angular sizes (S). The visual performance scale

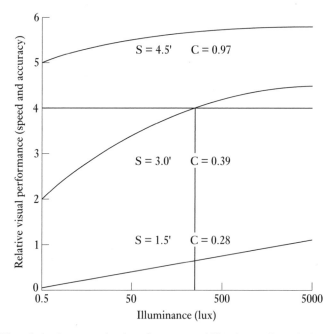

Fig. 1.6 The relation between visual performance and illuminance for task sizes and contrast

is relative and, if the task is very small and the contrast poor, there always seems to be room for improvement. If the task is large and the contrast good, then even moonlight could be sufficient to see by. This type of information is the basis for the recommended lighting levels quoted in codes and standards.

Example

A task comprises reading detail 0.5 mm wide at 0.5 m viewing distance. The reflectance of the task is 0.3 and the immediate background 0.2. What lighting level should ensure a relative visual performance of 4?

Solution

The contrast can be found from the reflectances

$$C = \frac{0.3 - 0.2}{0.2}$$

$$= 0.5$$

Figure 1.6 uses angular size in minutes of arc. In this case,

$$\text{Angular size} = \frac{0.5}{500}$$

$$= 0.001 \text{ rad}$$

$$= 0.001 \times \frac{180 \times 60}{\pi} \text{ minutes of arc}$$

$$= 3.43 \text{ minutes of arc}$$

This particular curve is not on Fig. 1.6 but can be approximately deduced. This will give an illuminance around 300 lx, although any accuracy is difficult without more detail in the figure.

VISUAL PREFERENCE

People normally have the right to express their own wish or preference as to how their living conditions should be. This is partly based on what people might expect to have in relation to those around them and partly on the minimum that they know they need. Motorists may prefer to have a good performance, top of the range family car because this would compare favourably with those around them. They know that a basic three-door hatchback would meet their needs. They would, however, consider a 3 litre luxury saloon car well beyond their needs.

The same range of choice can be applied to lighting. Figure 1.7 gives a summary of the results of different researchers when offering people a free choice of office lighting. Too many other factors are involved to draw conclusions beyond the fact that people prefer generous lighting standards, but this may not be sufficient justification to grant their wishes.

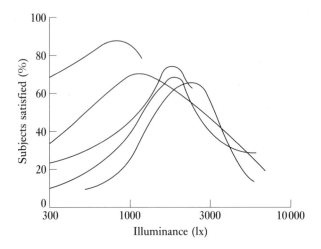

Fig. 1.7   Curves from various research reports on office workers' visual preference of illuminance at work (adopted from: CIE Publication No. 29, 1975)

EXERCISES

1.1   Write brief notes on the function and construction of the human retina. What is meant by the term 'presbyopia'?

1.2   Explain the term 'contrast' and how it may be used to assess the visibility of a specific task. How is this relevant to specifying suitable illuminance levels?

1.3   Two discharge lamps emit 20 per cent of their energy as radiation within the visible spectrum. Lamp A emits equal amounts at all wavelengths. Lamp B emits twice as much in the 500–600 nm band as in the 400–500 and 600–700 nm bands. How will their luminous efficacies compare? Assume that radiation is constant over each of the three bands.

CHAPTER 2

UNITS OF LIGHT

TOPICS COVERED

DEFINITIONS
Luminous flux
Luminous intensity

DISCUSSION OF UNITS
Luminous intensity
Luminous flux
Illuminance
Luminance
Calculation of luminous flux from intensity distribution
The large source

Various units have been introduced in Chapter 1. They will now be discussed in greater detail.

Light units can be described as *psychophysical* as they are a combination of human response and physical units of power. They stand on their own and cannot be directly related to other physical units.

There are only four units, so the system should be simple enough to use, but it is surprising how many people misquote the units, thereby displaying a lack of understanding.

DEFINITIONS

The following definitions are based in BS 4727 Part 4: *Glossary of terms particular to lighting and colour*. The units are in the International System as defined in British Standard 3763.

Luminous flux

Symbol: ϕ. Unit: *lumen*. The light *emitted by a source*, or received by a surface. The quantity is derived from radiant flux (power in watts) by evaluating the radiation in accordance with the relative luminous efficiency of the 'standard' eye (V_λ).

Lumen

Symbol: *lm*. The SI unit of luminous flux.

Luminous intensity

Symbol: *I*. Unit: *candela*. The quantity which describes the power of a source or illuminated surface to emit light in a given direction.

Notes

1. One may also speak of the mean intensity in a group of directions.
2. Only for a source of *uniform* intensity is it not required to specify direction.

Candela

Symbol: *cd*. The SI unit of luminous intensity, equal to 1 lumen per steradian.

Illuminance

Symbol: *E*. Unit: *lux*. The luminous flux density at a point on a surface, i.e. the luminous flux incident per unit area.

Lux

Symbol: *lx*. The unit of illuminance, equal to 1 lumen per square metre.

Luminance

Symbol: *L*. Unit: *candela per square metre*. The intensity of the light emitted in a given direction per projected area of a luminous or reflecting surface.

Note   Only for a source of *uniform* luminance is it not required to specify direction.

Candela per square metre

Symbol: *cd/m²*. The unit of luminance, equal to 1 candela per square metre of a projected area.

DISCUSSION OF UNITS

Light rays travel in a straight line. As the light rays will be diverging, the amount of light incident per unit area will decrease as the receiving surface moves away from the light source.

Figure 2.1 shows a light source emitting equally in all directions. It also shows that the relation between the flux and the uniform intensity in a solid angle is given by

$$\phi = I \times \omega \qquad\qquad [2.1]$$

Light is radiated in three dimensions even though, for convenience, data are recorded in terms of two dimensions. The surface of this paper is two dimensional: it has breadth and width. A distant point and the edges of the paper form a pyramid, a three-dimensional figure containing a given solid angle.

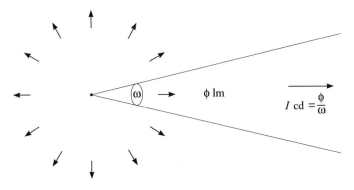

Fig. 2.1   Relation between flux and intensity for a small source

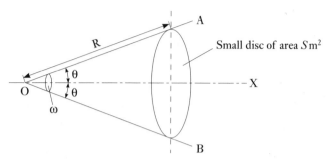

Fig. 2.2   A solid angle

In a similar fashion a plane angle can be drawn on a flat surface and can be referred to as a *planar angle*, measured in radians or degrees. A solid angle can only be drawn to suggest a solid and is measured in *steradians* (sr).

A solid angle is illustrated in Fig. 2.2.

Angle AOB is a plane angle 2θ. Spin angle AOB around OX and a cone is formed.

$$\text{Angle AOB (in radians)} = \frac{AB}{R} \quad \text{(provided angle AOB is small)}$$

$$\omega \text{ (solid angle)} = \frac{\pi(AB)^2}{4R^2} \text{ approx.} \tag{2.2}$$

If θ is 90° or $\pi/2$ rad, ω is 2π sr. If θ is 180° or π rad, ω is 4π sr.

Example

If, in Fig. 2.2, θ is 10°, what is the value of ω?

Solution

$$\text{If } \theta = 10° \text{ then } \frac{AB}{2} = R \sin\theta$$

$$\therefore \quad AB = 2R \sin 10°$$

Substituting in eqn [2.2],

$$\omega = \frac{\pi(2R \sin 10°)^2}{4R^2}$$

$$= 0.095 \text{ sr}$$

Luminous intensity

The concept and definition of 'luminous intensity' may be difficult to grasp. Figure 2.1 shows luminous flux as being emitted within a solid angle, which suggests a cone of substantial size. On the other hand, the definition of luminous intensity considers a cone of such small dimensions that it becomes a single line, i.e. a specific single direction.

Flux in a solid angle is the product of the mean luminous intensity and a solid angle. If the solid angle becomes so small that it is associated with a single direction – i.e. if the solid angle approaches zero – it is difficult to conceive that there is any flux at all! It is more helpful if we do not think of the solid angle as being infinitely small, but as being just small enough that light may be considered to be emitted evenly within it. Luminous intensity is the unit used to define the 'strength' of light in specific directions, and from that unit we can derive the other three, as follows.

Luminous flux

This is derived from eqn [2.1] and is dimensionally the same as intensity.

Illuminance

This is luminous flux falling on a unit area of surface and, hence, is related to intensity.

Luminance

Equation [2.1] does not define the size of the source emitting lumens. If it is a flat area S m^2, then

$$L = \frac{I}{S_A} \, \text{cd/m}^2 \qquad\qquad [2.3]$$

It is important to appreciate that S_A is the area *as seen* from the direction specified, called the *apparent* or *projected* area. Figure 2.3 shows some examples of apparent areas.

Example

A sphere of diameter 0.5 m emits 2000 lm uniformly in all directions. Calculate the average intensity, luminance and illuminance on a surface 3 m from its centre.

Solution

The average intensity is derived from eqn [2.1]. In this case the total solid angle 4π is taken since the emission of light is the same in all directions:

$$\phi = I \times \omega \, \text{lm}$$

$$\therefore \quad I = \frac{\phi}{\omega}$$

$$= \frac{2000}{4\pi} \, \text{cd}$$

$$= 159 \, \text{cd}$$

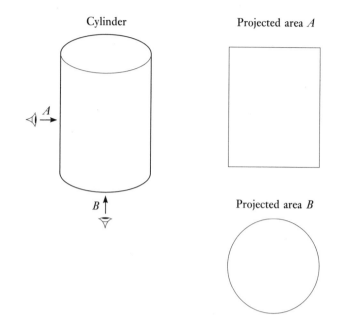

Cylinder   Projected area *A*

B

Projected area *B*

Fig. 2.3   Projected areas of a cylinder

Self-assessment task 2.1

What is the basic difference between luminous flux and luminous intensity?
    If two lamps have the same flux output do they necessarily have the same intensity distribution?

The luminance is derived from eqn [2.3]:

$$L = \frac{I}{S_A} \text{ cd/m}^2$$

$$= \frac{159}{\text{apparent area of sphere}} \text{ cd/m}^2$$

$$= \frac{159}{\pi \times 0.25^2} \text{ cd/m}^2$$

$$= 810 \text{ cd/m}^2$$

The illuminance requires a development of eqn [2.2]. Referring to Fig. 2.2, and finding the illumination on the small disc,

$$E = \frac{\text{lumens received on disc}}{\text{area of disc}} \text{ lx}$$

$$= \frac{I \times \omega}{\pi (AB)^2/4}$$

From eqn [2.2]

$$\omega = \frac{\pi(AB)^2}{4 \times (\text{distance})^2}$$

The distance is from the centre of the sphere to the measuring surface which, in this case, will be 3 m.

Substituting,

$$E = \frac{I}{\pi(AB)^2/4} \times \frac{\pi(AB)^2/4}{(\text{distance})^2}$$

$$= \frac{I}{(\text{distance})^2} \text{ lx} \qquad\qquad [2.4]$$

(this equation is known as the *inverse square law of illumination*)

$$= \frac{159}{9} \text{ lx}$$

$$= 17.7 \text{ lx}$$

Luminous flux

The lumen is used to define the light output of lamps, e.g. a 60 watt GLS lamp emits 660 lm. It is also used to quote the luminous efficacy of lamps (see Chapter 1). Typical luminous efficacies are given in Table 2.1.

If the lumens in a specific zone are required – e.g. the upward lumens from a luminaire – then the calculation involves further knowledge of intensity. This will be considered in the next section.

Table 2.1 Typical luminous efficacies

Light source	Average efficacy (lm / W)
Low-pressure sodium lamp	200
High-pressure sodium lamp	110
Sunlight	80
White fluorescent tube	90
Tungsten–halogen lamp	22
GLS filament lamp	15

Luminous intensity

Unless referring to mean or average intensity, this unit is always related to direction. Direction must normally be specified in terms of three axes. This raises problems when using flat diagrams as these can only show two of the axes. Plots of intensity distribution are shown on a flat diagram (polar coordinate graph) and this will refer to the pattern of intensity distribution in a specific plane.

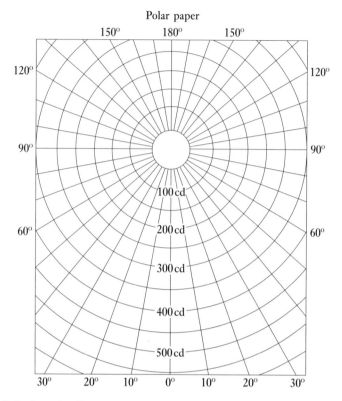

Polar paper

Fig. 2.4 Polar intensity diagram

Intensity distribution diagrams (polar curves)

The two variable dimensions are intensity (candelas) and direction (degrees). The simplest diagram for displaying this information is the polar curve (see Fig. 2.4), which is used to show the intensity distribution, usually in a vertical plane. The 'spokes' represent the various angles from the downward vertical and the circles the magnitudes of intensity. Figure 2.5 shows a typical curve for a half-plane, i.e. where the intensity distribution is common to all planes. It is only necessary to produce a half-plane polar curve. The point of origin of the angles is the photometric centre of the light source.

Example

The following table gives a list of intensities at different vertical angles. Plot the polar curve.

Angle from downward vertical (°)	0	15	30	45	60	75	90
Intensity (cd)	620	570	400	200	50	10	0

Solution
This is shown on Fig. 2.5.

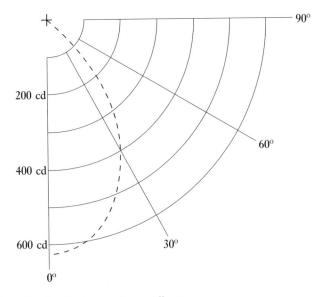

Fig. 2.5   Intensity distribution – polar coordinates

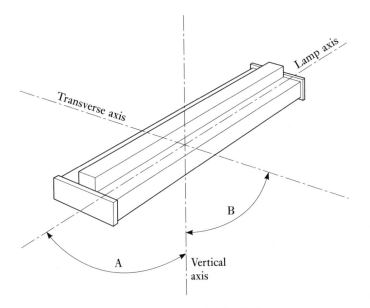

Fig. 2.6   Major vertical planes (courtesy Philips Lighting Ltd)

It is important to appreciate that this curve is for a specific vertical plane. In the case of a fluorescent luminaire, the plane containing the lamp axis usually has a different set of values of intensity to those obtained in the plane across the lamp axis. These axes are shown in Fig. 2.6.

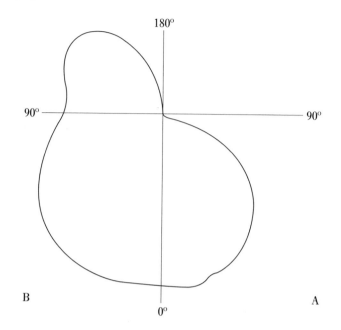

Fig. 2.7   Polar curve for general diffusing fluorescent luminaire

In this case the light distribution around the luminaire would be called *asymmetric*. If the polar curves for all the vertical planes are nearly the same, then the distribution is *symmetric*.

Some manufacturers include a schematic illustration of the polar curve. The shape gives an indication of the light distribution and an example for a surface-mounted linear diffuser is shown in Fig. 2.7.

For asymmetric distributions the convention is to have the transverse plane (B in Fig. 2.6) on the left and the axial plane (A in Fig. 2.6) on the right.

In the 1980s a system called British Zonal Classification (BZ) was used but this is now obsolete. Intensity distribution diagrams are normally based on 1000 lamp lumen output.

Cartesian coordinates

In some cases, such as for floodlights, it is perhaps clearer to display the intensity distribution on rectangular (Cartesian) coordinates. The bottom axis is the angle (*x* axis) and the vertical axis is the intensity (*y* axis), as shown in Fig. 2.8.

Iso-candela diagram

The intensity distributions of some luminaires (e.g. road lighting lanterns) are so complex that a large number of plane polar curves would be required to record such a distribution.

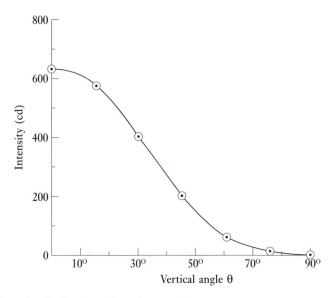

Fig. 2.8   Intensity distribution – Cartesian coordinates

Complex distributions are displayed on a web called an iso–candela diagram. With such a diagram, it is possible to display the intensities in all directions in a hemisphere.

Figure 2.9 shows such a web, which is similar to the web used by geographers to display the height contours of the earth's surface. Visualise the diagram as not flat, but hemispherical and curving upwards from the paper. The curves drawn upon it, roughly vertically and horizontally, correspond to the lines of longitude and latitude on a globe.

Imagine that the luminaire is miniaturised and is located at the centre of the globe and that the surface of the globe is transparent, apart from the lines of longitude and latitude.

Mark on the surface of the globe all those points where the intensity has the same value; for example, all planes where the value is 100 cd can be dotted and the dots joined to give a contour line, just like the equal height contours on a map. Similarly, all dots representing 200 cd can be joined – and so on – for as many contour lines as are needed to illustrate the distribution of intensity.

Self-assessment task 2.2

Look at the iso–candela diagram, Fig. 2.9. Mark in pencil the position of the luminaire. Calculate the approximate average intensity 70° below the horizontal plane through the luminaire.

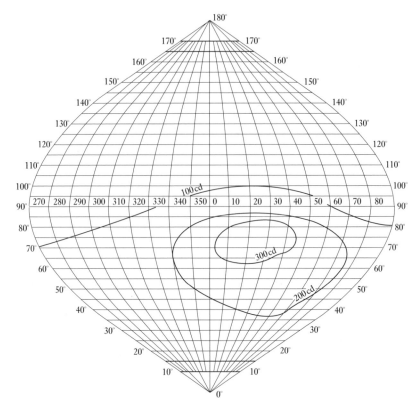

Fig. 2.9   Iso-candela diagram

CIE classification

This is a simpler classification, based not on the intensity distribution, but in terms of the total luminous flux directed above and below the horizontal, and shown in Table 2.2.

Table 2.2   CIE classification of luminaire photometric distribution

Luminaire	Flux above horizontal (%)	Flux below horizontal (%)
Direct	0–10	90–100
Semi-direct	10–40	60–90
General diffuse	40–60	40–60
Semi-indirect	60–90	10–40
Indirect	90–100	0–10

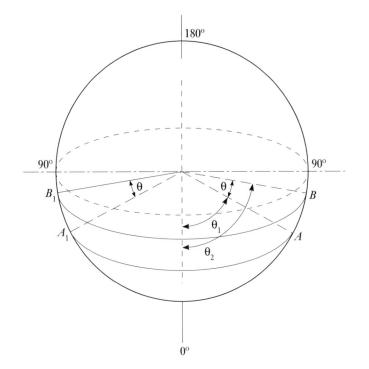

Fig. 2.10 Relation between plane and solid angle (zone factors)

Calculation of luminous flux from intensity distribution

The relationship $\phi = I \times \omega$ (eqn [2.1]) has already been discussed. If the intensity distribution represents the average of all vertical planes then the information of the polar curve is sufficient to calculate the flux in any zone bounded by two horizontal planes. This is shown in Fig. 2.10. The plane angle θ has to be converted to its equivalent solid angle ω.

The normal transformation is given by

$$\omega = 2\pi(\cos\theta_1 - \cos\theta_2) \qquad [2.5]$$

where θ_1 and θ_2 are the angles, as shown on Fig. 2.10.

This expression is referred to as the *zone factor* (ZF) and Table 2.3 shows the zone factors if the plane angle zones are at $10°$ intervals. The values from $90°$ to $180°$ will be the mirror image of those from $90°$ to $0°$, i.e. $ZF(80° - 90°) = 1.091 = ZF(100° - 90°)$.

Having now established a relationship between plane angle (θ) and solid angle (ω) for $10°$ zones, the mid-intensity values are determined for each zone, and hence the luminous flux in that zone, from

Table 2.3   Zone factors

Vertical angle θ (°)	Zone factor (ω)	Vertical angle θ (°)	Zone factor (ω)
0–10	0.095	50–60	0.897
10–20	0.283	60–70	0.993
20–30	0.463	70–80	1.058
30–40	0.628	80–90	1.091
40–50	0.774		

$$\text{Luminous flux in zone } \theta_1 \text{ to } \theta_2 = \frac{I_{\theta_1} + I_{\theta_2}}{2} \times 2\pi(\cos\theta_1 - \cos\theta_2) \qquad [2.6]$$

Using zone factors it is possible to calculate the efficiency of luminaires quoted in terms of light output ratio:

$$\text{Light output ratio (LOR)} = \frac{\text{total lumen output of luminaire}}{\text{total lamp lumen output}}$$

$$\text{Downward light output ratio (DLOR)} = \frac{\text{downward lumen output of luminaire}}{\text{total lamp lumen output}}$$

$$\text{Upward light output ratio (ULOR)} = \frac{\text{upward lumen output of luminaire}}{\text{total lamp lumen output}}$$

$$\text{LOR} = \text{DLOR} + \text{ULOR}$$

Example

A luminaire has an intensity distribution of $100 \cos\theta$ cd and emits no light above 90°. If the lamp emits 500 lm, what is the LOR of the luminaire?

Solution

This is best dealt with by constructing a table (Table 2.4).

$$\text{Total light output} = \text{sum of column D}$$
$$= 312 \text{ lm}$$
$$\text{LOR} = \frac{312}{500}$$
$$= 0.62$$

In this case, DLOR is also 0.62. As there is no light above 90°, ULOR is 0.

Illuminance

Deducing illuminance is the most common form of calculation that a lighting engineer has to perform. It is the most precise part of lighting specification and is the only calculation that can be checked, with any reasonable degree of accuracy, in a completed installation. There are two basic situations:

Table 2.4

A Zone (°) $\theta_1 - \theta_2$	B Average intensity (cd) $I_0 \cos[(\theta_1 + \theta_2)/2]$	C Zone factor (sr) $2\pi(\cos \theta_1 - \cos \theta_2)$	D Flux (lm) B × C
0–10	99.6	0.095	9.5
10–20	96.6	0.283	24.4
20–30	90.6	0.463	42.0
30–40	81.9	0.628	51.4
40–50	70.7	0.774	54.8
50–60	57.4	0.897	51.0
60–70	42.3	0.993	42.0
70–80	25.9	1.058	27.4
80–90	8.7	1.091	9.5
90–100	0	6.28	0

1. the direct illuminance at a point on a specified plane (E_D);
2. the average illuminance on a room surface due both to direct flux and that received by reflection off other surfaces (E_{av}).

The first situation will be considered in this chapter, the second in Chapter 7.

Calculation of direct illuminance

The point source   This is an important concept. In theory a point source is infinitely small, but in practice it is a light source small enough to consider that all the light comes from a clearly defined point in space. This considerably simplifies the calculations and it is reasonable to assume that a light source is a point if the measuring distance is at least five times the maximum luminous width of the source. For instance, for a 1.5 m long fluorescent tube, the measuring distance would have to be at least 7.5 m (which is a far greater distance than normally experienced in a situation where illuminance calculations are needed).

The inverse square 'law'   Referring to Fig. 2.11, the illuminance at a point A on a horizontal plane directly below the source is

$$E = \frac{I_0}{H^2} \, \text{lx} \qquad\qquad [2.7]$$

where I_0 is the intensity towards point A (cd) and H is the height of the source above the plane (m).

Now considering point B, three things have changed:

1. The measuring distance is now PB.
2. The intensity is now I_θ.

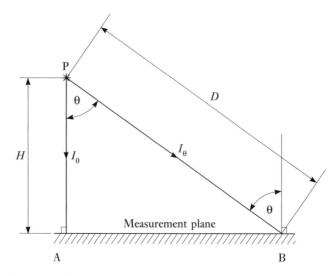

Fig. 2.11   Illuminance due to a small light source

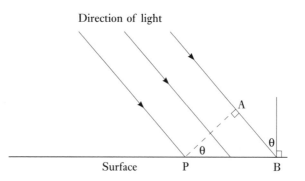

Fig. 2.12   Tilting the plane of illuminance – cosine effect

3. The surface is no longer at right angles or normal to the direction of the
   incident light; it is effectively tilted away from the normal by the angle θ and
   Fig. 2.12 indicates that the effective area has been increased by the ratio PB
   to PA, where

$$\cos \theta = \frac{PA}{PB}$$

Hence, the illuminance falls by the factor of $\cos \theta$, and the illuminance at B is

$$E = \frac{I_\theta}{PB^2} \cos \theta \ \text{lx}$$

[2.8]

This can be simplified to two variables:

$$\cos \theta = \frac{H}{PB}$$

$$E = \frac{I_\theta}{H^2} \cos^3 \theta \ \text{lx} \qquad [2.9]$$

This is sometimes called the $\cos^3$ 'law' of illumination.

Self-assessment task 2.3

Which lighting units are related by the inverse square 'law'? Why should the cosine effect reduce the calculated illuminance value when a surface is at an angle to the incident light?

Example

A small luminaire has the following intensity distribution:

$\theta(°)$	0	20	40	60	80	90
I_θ (cd)	420	450	380	200	70	0

It is mounted 2.5 m above the working plane. How will the illuminance vary on this plane?

Solution
The first stage is to plot the polar curve. The illuminance at any point on the plane is found from eqn [2.9]. The results can be set out as in Table 2.5.

Isolux diagrams   It can be helpful to display the pattern of illuminance on a specific plane as a series of illuminance contours. Figure 2.13 is an isolux diagram for the data used in the previous example. This type of diagram does provide a method for planning the illumination of a large area such as a car park, its main use being for exterior lighting where there is no reflected light to consider.

Beam or cone illuminance diagrams   A convenient display of direct illuminance at various mounting heights by symmetrical luminaires such as recessed spotlights is the beam diagram (also referred to as the cone diagram).

Table 2.5

θ (°)	I_θ (cd)	$\cos \theta$	E (lx)	X (m)
0	420	1	67.2	0
20	450	0.830	59.8	0.9
40	380	0.450	27.4	2.1
60	200	0.125	4.0	4.3
80	70	0.040	0.5	6.9
90	0	0	0	–

X is obtained from $X = H \tan \theta$. E can then be plotted against X.

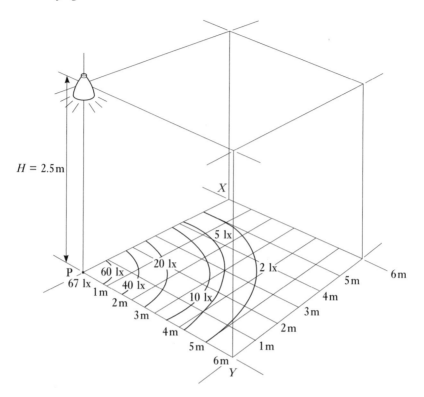

Fig. 2.13   Isolux diagram

Figure 2.14 illustrates an example for a range of reflector lamps. The vertical scale indicates the height to the plane of measurement and the peak illuminances at the beam centre are given at the side; 3 m directly below the 100 W lamp the luminance is 165 lx. Unless otherwise stated, these values refer to the lamp initial lumen output, they are not usually based on 1000 lm requiring correction.

There may be two beams as shown in Fig. 2.14. The inner beam is for a maximum : minimum intensity ratio of 2:1 and a beam angle of 30°. The larger beam is for a ratio of 10:1 and a beam angle of 80°.

To plan an even illuminance the smaller beam is used. However, for display lighting a greater variation can be accepted and the larger beams can be used.

Although beam diagrams refer to illuminance, the beam angles are often based on intensity. For angles below 30° this only gives small errors, but for wider angles the manufacturer should be consulted.

Example

A 100 W lamp is mounted 3 m above the floor along a corridor. Calculate the spacing and illuminance below a row of lamps to achieve a uniformity ratio of 5:1.

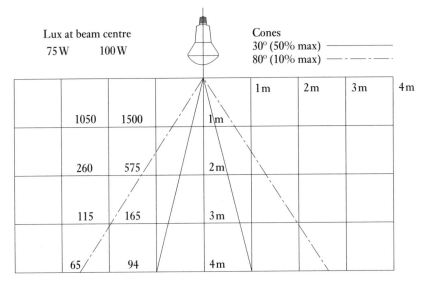

Lux at beam centre		Cones
75 W	100 W	30° (50% max) ————
		80° (10% max) — · — · — ·

				1 m	2 m	3 m	4 m
1050	1500		1 m				
260	575		2 m				
115	165		3 m				
65	94		4 m				

Fig. 2.14 Beam diagram

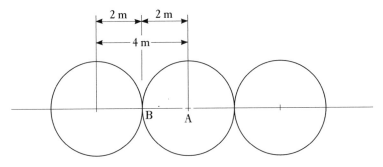

Fig. 2.15 Spacing of 10 per cent beams to produce 5:1 uniformity

Solution

If the 10 per cent beams are drawn to touch each other as in Fig. 2.15, this uniformity will be achieved.

■ The illuminance at A will be 165 lx.
■ The illuminance at B will be 2 × 16.5 lx, or 33 lx.

This calculation only gives the direct illuminance value and does not include light reflected off the room surfaces. Nor has it made any allowance for a light loss factor. Illuminance values at other heights can be calculated using the inverse square law.

For example, the maximum value at 2.5 m is

$$E = 165 \times \frac{3^2}{2.5^2}$$

$$= 237.6 \text{ lx}$$

The large source

When the source size becomes significant relative to the distance, it is no longer possible to identify the specific distance or intensity. A new expression is needed which allows for the size of the source. The method of deriving these expressions is to consider a small element which obeys the inverse square law and then to integrate this expression for the whole source.

In arriving at these expressions it is assumed that the source has a uniform diffusing surface.

The disc source

To find the illuminance on a parallel plane   Figure 2.16 shows a uniformly diffusing disc. The illuminance at a point P directly below the centre is

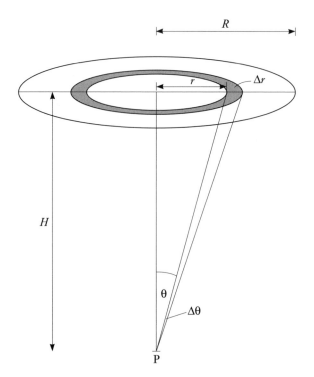

Fig. 2.16   Illuminance from a disc source

$$E = \pi L \frac{R^2}{(R^2 + H^2)} \text{ lx} \qquad [2.10]$$

where L is the luminance (cd/m^2), R is the radius and H is the distance.

Derivation Consider a small ring or radius r and width Δr. This will obey the inverse square law since all parts are equidistant from P. If a diffusing surface has a luminance of L cd/m^2 then, from eqn [2.3], the intensity in a specified direction is

$I = L \times$ projected area cd

e.g. for a flat luminous panel of area S m^2 and luminance L cd/m^2 when viewed at angle θ

$I = L \times S \cos \theta$ cd

Considering the elemental ring and the illuminance at P,

$$E = \frac{I}{\text{distance}^2} \times \cos \theta \text{ lx}$$

To find I:

The area of ring $= \Delta r \times 2\pi r$ m^2
$$\therefore \quad I = \Delta r \times 2\pi r \times \cos \theta L \text{ cd}$$
$$\text{Distance} = H \sec \theta \text{ m}$$
$$\therefore \quad E = \frac{2\pi r \Delta r \times \cos \theta \times \cos \theta \times L}{H^2 \sec^2 \theta} \text{ lx}$$

This expression contains two variables, θ and r. To integrate put r in terms of θ:

$$\tan \theta = \frac{r}{H}$$
$$\therefore \quad \sec^2 \theta \, d\theta = \frac{dr}{H}$$

Taking the limiting condition $\Delta r \to dr$ and substituting r for θ,

$$E = \frac{2\pi L}{H^2} r \, dr \text{ lx}$$
$$= \frac{2\pi L}{H^2} H \tan \theta \, H \sec^2 \theta \cos^4 \theta \, d\theta \text{ lx}$$
$$= 2\pi L \sin \theta \cos \theta \, d\theta \text{ lx}$$
$$\therefore \quad \text{Total illuminance} = 2\pi L \int_{\theta=0}^{\theta=\tan^{-1}(R/H)} \sin \theta \cos \theta \, d\theta \text{ lx}$$
$$= 2\pi L \left[\frac{\sin^2 \theta}{2} \right]_{\theta=0}^{v=\tan^{-1}(R/H)} \text{ lx}$$
$$= \pi L \frac{R^2}{R^2 + H^2} \text{ lx}$$

Alternatively, this expression could be written

$$E = \pi L \sin^2 \alpha \text{ lx}$$

where α is half the angle subtended by the source at P; or

$$E = \pi L \text{ (function } \alpha) \text{ lx}$$

This function of α can be termed the aspect factor (AF) as it is a function of the aspect angle. A table can be provided of (function α) against α and this can be used to simplify calculations and to consider surfaces other than uniform diffusing.

Example

If a disc has a luminance of 400 cd/m² and a radius of 1.5 m, what is the illuminance on a parallel plane 2 m below?

Solution

$$E = \pi \times 400 \times \frac{1.5^2}{1.5^2 + 2^2} \text{ lx}$$

$$= 452 \text{ lx}$$

Approximations   It is possible to use this last formula for other shaped-area sources. There is a degree of error, but it need not be significant.

Self-assessment task 2.4

Looking at the example above, calculate the degree of error if the disc is treated as a point source, i.e. formula [2.7] is used.

Example

Calculate the illuminance 2 m below the centre of a 3 m square luminous ceiling which emits 10 000 lm, given the relationship that

$$\text{Luminance} = \frac{\text{lumens emitted per m}^2}{\pi}$$

Solution
The area of an equivalent disc is

$$\pi R^2 = 3 \times 3 \text{ m}^2$$

$$\therefore \quad R^2 = \frac{9}{\pi} \text{m}^2$$

$$\therefore \quad E = \pi \times \frac{10\,000}{9} \times \frac{9/\pi}{[9/\pi + 2^2]}$$

$$= 1455 \text{ lx}$$

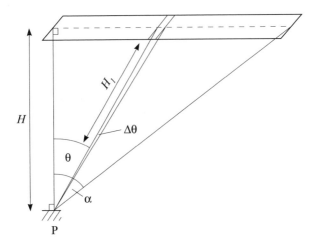

Fig. 2.17 Illuminance from a linear strip source on a parallel plane

The linear source

In similar fashion, and using the notation in Fig. 2.17, the illuminance on a parallel plane directly below one end of a uniform diffusing strip source is given by

$$E = \frac{I_0}{2H}(\alpha + \sin \alpha \cos \alpha) \text{ lx} \qquad [2.11]$$

where I_0 is the downward intensity per metre length of source and α is the aspect angle in radians.

This expression can be derived in a similar way to the disc, but in this case the element is a small transverse strip as shown in Fig. 2.17. It is obtained from the expression

$$E = \frac{I_0}{H_1}\int \cos^2 \theta \, d\theta \text{ lx}$$

where the limiting value of θ is α, the aspect angle.

The value of E is opposite one end. Where the point lies between the ends, or beyond, the source must be split into separate elements as in Fig. 2.18(a) and (b).

If the point of measurement is to one side, as in Fig. 2.18(c), then the formula is modified to

$$E = \frac{I_\theta}{2H_1}(\alpha + \sin \alpha \cos \alpha) \cos \theta \text{ lx} \qquad [2.12]$$

where α is the new aspect angle, and θ is the angular displacement in the transverse plane. If the source is a flat diffusing strip, then

$$I_\theta = I_0 \cos \theta$$

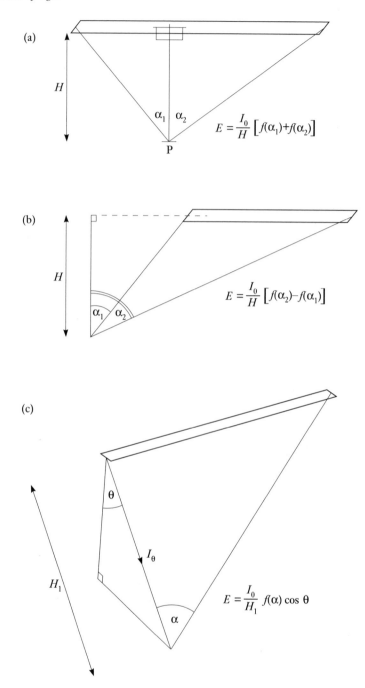

(a) H α_1 α_2 $E = \dfrac{I_0}{H}\left[f(\alpha_1)+f(\alpha_2)\right]$

P

(b) H α_1 α_2 $E = \dfrac{I_0}{H}\left[f(\alpha_2)-f(\alpha_1)\right]$

(c) θ I_θ H_1 α $E = \dfrac{I_0}{H_1}\,f(\alpha)\cos\theta$

Fig. 2.18   Illuminance from a linear strip source on a parallel plane

and the expression would become

$$E = \frac{I_\theta}{2H}(\alpha + \sin \alpha \cos \alpha) \cos^2 \theta \text{ lx} \qquad [2.13]$$

Example

A fluorescent lamp is mounted directly above the front edge of a workbench. Both are 1.5 m long and the mounting height is 1 m. The transverse intensity is 450 cd at all angles. Calculate the illuminance (a) below one end of the bench and (b) in the centre.

Solution

(a) The illuminance below one end can be found as follows:

$$I_0 = \frac{450}{1.5} \text{ cd/m}$$

$$= 300 \text{ cd/m}$$

$$\tan \alpha = \frac{1.5}{1.0}$$

Hence

$$\alpha = 0.98 \text{ rad}$$

$$E = \frac{300}{2 \times 1}(0.98 + 0.83 \times 0.56) \text{ lx}$$

$$= 217 \text{ lx}$$

(b) To find the illuminance in the centre above the front edge, there are two equal calculations. Hence

$$E = 2 \times \frac{300}{2 \times 1}(0.64 + 0.60 \times 0.80) \text{ lx}$$

$$= 337 \text{ lx}$$

The above calculations refer to a parallel plane.
  If the plane is at right angles, as in Fig. 2.19, then

$$E = \frac{I_0}{2H} \sin^2 \alpha \text{ lx} \qquad [2.14]$$

Again the calculation refers to a point opposite one end, but this is a far less common calculation.

Use of aspect factor with linear sources   Equation [2.11] can be modified to

$$E = \frac{I_0}{wH}(\text{function } \alpha)$$

where w is the width of the source.

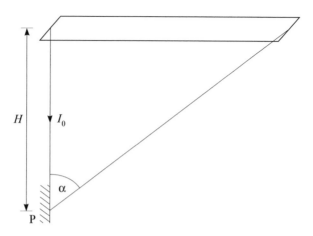

Fig. 2.19   Illuminance from a linear strip source on a perpendicular plane

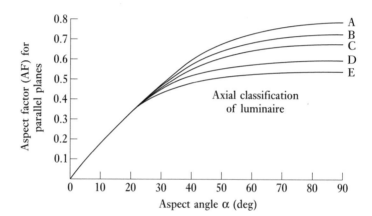

Fig. 2.20   Aspect factor for linear source

Since $I_0 = L \times$ area
$I_0/\text{m} = L \times$ width $\times 1$

$$\therefore \quad \frac{E}{L} = K \text{ (function } \alpha)$$ [2.15]

This can be taken a stage further, since although the formula was for a uniformly diffusing strip where $I_\theta = I_0 \cos \theta$, it is extended to include other axial distributions.

Figure 2.20 shows a range of aspect factors for different axial intensity distributions. Figure 2.21 shows how a particular axial distribution can be classified. In the case of the examples, the axial distribution is cosine – i.e. that of a uniform diffusing surface – and the classification is A. If there is control of the axial distribution, such as in a prismatic enclosure, the classification will lie between B and E.

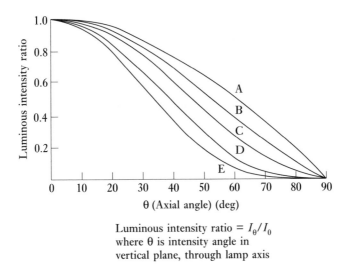

Luminous intensity ratio = I_θ/I_0
where θ is intensity angle in
vertical plane, through lamp axis

Fig. 2.21   Classification of axial distribution for use in aspect factor

Example

Taking the previous example, find the illumination at the far corner of the bench if it is
1 m wide.

Solution

The aspect angle α is given by

$$\alpha = \tan^{-1}(1.5/\sqrt{2})$$
$$= 46.7°$$

From Fig. 2.18 for classification A, the aspect factor (AF) is 0.66.

$$\therefore \quad E = \frac{I}{lH_1} \times AF \times \cos\theta \text{ lx}$$

$$= \frac{450}{1.5 \times \sqrt{2}} \times 0.66 \times \cos 45° \text{ lx}$$

$$= 98.6 \text{ lx}$$

Checking, by using the original formula,

$$E = \frac{I}{l} \times \frac{1}{2H_1}(\alpha + \sin\alpha\cos\alpha)\cos\theta \text{ lx}$$

$$= \frac{450}{1.5} \times \frac{1}{2 \times \sqrt{2}}(0.815 + 0.728 \times 0.686)\cos 45° \text{ lx}$$

$$= 98.6 \text{ lx}$$

The rectangular source

The formulae for rectangular source calculations are not easy to derive and these problems can be solved using tables. The main use of this type of calculation is for windows as these are normally considered to be rectangular diffusing light sources. If the calculation is for side windows, then the plane of illumination is perpendicular to the source; for roof lights the planes will be parallel.

Illuminance for non-planar surfaces

The preceding sections have dealt with the calculation of direct illuminance at a point on a flat (or planar) surface. There are a number of situations where this information may be misleading. For example, the illumination of a human face involves vertical and three-dimensional surfaces. The horizontal planar value has little relevance.

Two alternatives at present being considered are the spherical surface (scalar illuminance), and the all-round vertical surface (cylindrical illuminance). Methods of calculating and using these values will be found in Chapter 7.

Luminance

This unit has already been partially introduced, it being impossible to discuss other units without reference to it. It expresses the light emitted per unit projected area of light source or reflecting surface. The unit is the candela per square metre (cd/m^2).

There is an alternative non-SI unit, the apostilb (asb). This expresses the light emitted in terms of lumens per unit actual area. It is an easier unit to comprehend since

Luminance (asb) = illuminance (lx) × reflectance

There is a simple relationship between the two units:

Luminance (asb) = π × luminance (cd/m^2) [2.16]

Concept of perfect diffusing surface

Calculations are usually simpler if reflecting and emitting surfaces can be considered perfectly diffusing. This means the luminance is the same in *all* directions.

Consider a flat diffusing surface of area S m^2, at $\hat{\theta}$:

$$L = \frac{I_\theta}{\text{projected area}} \text{ cd/m}^2$$

$$= \frac{I_\theta}{S \cos \theta} \text{ cd/m}^2$$

at $\hat{\theta}$:

$$L = \frac{I_0}{S} \, \text{cd/m}^2$$

$$\therefore \quad \frac{I_\theta}{S \cos \theta} = \frac{I_0}{S}$$

since L is constant. Therefore,

$$I_\theta = I_0 \cos \theta$$

This surface can be referred to as a 'cosine diffuser'.

Relationship between lumens and intensity for flat diffuser

In Fig. 2.22 consider a hemispherical shell around a surface S. If S emits ϕ lumens, all these lumens will land on the shell surface.

Consider a small ring of width $R \, \Delta\theta$ and radius r:

$$\text{Area of ring} = \text{width} \times \text{circumference}$$
$$= R \Delta\theta \times 2\pi r$$
$$= R \, \Delta\theta \times 2\pi R \sin \theta$$
$$\therefore \quad \text{Area of ring} = 2\pi R \sin \theta \times R \, \Delta\theta$$

$$\text{Illuminance on ring} = \frac{I_\theta}{R^2} \, \text{lx}$$

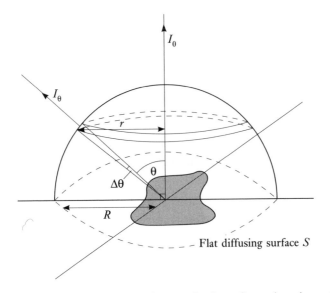

Fig. 2.22 Flat diffusing surface – relation between luminous flux and maximum intensity

∴   Lumens on ring = illuminance × area

$$= \frac{I_\theta}{R^2} 2\pi R^2 \sin\theta \Delta\theta \text{ lm}$$

Lumens on hemisphere $= 2\pi \int I_\theta \sin\theta \, d\theta \text{ lm}$

But, as $I_\theta = I_0 \cos\theta$,

Lumens on hemisphere $= 2\pi I_0 \int \cos\theta \sin\theta \, d\theta \text{ lm}$

$$= 2\pi I_0 \left[\frac{\sin^2\theta}{2} \right]_{\theta=0}^{\theta=\pi/2} \text{ lm}$$

Applying the limits,

$$\phi = \pi I_0 \text{ lm} \tag{2.17}$$

If both sides of [2.17] are divided by the area, then

$$\frac{\phi}{S} = \frac{\pi I_0}{S}$$

or

Luminance (abs) $= \pi \times$ luminance (cd/m²)

which proves eqn [2.16].

There are three important relationships:

Flat diffuser:	lumens = $\pi \times$ maximum intensity	[2.18]
Spherical diffuser:	lumens = $4\pi \times$ average intensity	[2.19]
Cylindrical diffuser:	lumens = $\pi^2 \times$ maximum intensity	[2.20]

Self-assessment task 2.5

Give examples of luminaires for each of the equations [2.18], [2.19] and [2.20].

Example

A 1.5 m fluorescent tube has a diameter of 0.038 m and emits 5000 lm. What is its luminance?

Solution

Using eqn [2.20] the maximum intensity is

$$I = \frac{\text{lumens}}{\pi^2} \text{ cd}$$

$$= \frac{5000}{\pi^2} \text{ cd}$$

$$= 506.7 \text{ cd}$$

The projected area when viewing a tube sideways is that of a rectangle of 1.5×0.038 m^2:

$$\therefore \quad L = \frac{506.7}{1.5 \times 0.038} \; \text{cd/m}^2$$

$$= 8889 \; \text{cd/m}^2$$

This chapter has covered the basic calculations of direct luminous flux and illuminance. Calculations involving the reflected components are considered in Chapter 7.

EXERCISES

2.1   A small light source emits 1200 lm and is placed at the centre of an opal sphere of 0.5 m diameter. The opal glass can be considered to be of uniform luminance.

If 25 per cent of the lamp flux is absorbed, calculate the external luminance of the sphere.

What illuminance will the sphere produce on a surface 3 m below its centre?

2.2   Give definitions of the following:

(a) luminous flux
(b) luminous efficacy
(c) the lux
(d) a uniformly diffusing surface

Suspended 1.5 m vertically above each corner of a square horizontal table of sides 1 m is a lamp of uniform luminous intensity 100 cd. Calculate the illuminance at the corners of the table.

2.3   A 58 W fluorescent lamp, 1.5 m long, emits 5000 lm. It is shielded by a pelmet and is 0.3 m away and parallel to a wall. Calculate the illuminance opposite one end of the tube on the wall and at a second point on the wall 1 m lower down.

Would this arrangement be suitable for lighting a picture hung in this space?

2.4   Comparing the lamp in 2.3 which has a diameter of 26 mm with a lamp of the same length and lumen output, but a diameter of 12 mm, calculate their relative luminances.

What is the advantage in reducing tube diameter?

CHAPTER 3

COLOUR

TOPICS COVERED

MIXING OF COLOUR
   Additive
   Subtractive
   Specification of colour appearance of surfaces
   Specification of colour appearance of light sources
   Colour rendering of light sources
   Colour constancy
   Colour matching

This chapter deals with a subjective effect. The word 'colour' is meaningless to a blind person and it is difficult to explain in text a subject which can only properly be described by personal expression and comparison. The fact that natural adjectives are often used to describe colour illustrates the basic difficulty in attempting to specify colour in technical terms. Even when the technical terms are understood, the description 'blood red' gives an inaccurate but more acceptable impression of a red colour than 'Munsell R 4/14' or CIE $0.6x$, $0.3y$.

Coupled with the difficulty in description is the difficulty in expressing the subjective effect. In Chapter 1 the process of colour vision was outlined briefly, and if the energy of different wavelengths produces in the brain the sensation of different colours, then from Table 1.1 it can be deduced that green is a more 'efficient' colour than red in terms of lumens of light. However, red is an arresting colour used to signify danger and green is a 'safe' colour with less visual impact. This contradiction can only be explained by accepting that colours have a psychological as well as a physical effect. It is hard to anticipate the effect colour has on a lighting scheme, and dissatisfaction may arise with a lighting installation, not because there is insufficient light, but because the colour effect of the space or of the room surfaces is disliked.

There are some generalisations relating colour to the lighting of interiors. Most lamps produce some form of white light from cool to warm. This is their *colour appearance* and it is reasonable to say that the cooler appearance is associated with high lighting levels and working conditions. A warmer appearance is more for relaxation and socialising. The lamp industry provides the range and the designer has to make the choice.

Lamps also provide a wide range of *colour rendering* properties, from excellent to poor. The designer must decide how important this is. The room surface colours can have a dominant effect, not only on the appearance, but also on the lighting efficiency. Strong colours can play havoc with reflectances and lighting calculations.

Self-assessment task 3.1

On a scale of 1 to 5 rate the importance of colour appearance and colour rendering in the following locations:

(a) a hospital ward
(b) a general office
(c) a frozen food shop
(d) a general factory

MIXING OF COLOUR

Additive

Isaac Newton first demonstrated and explained the composition of white light, by refracting it through a glass prism into its constituent spectral colours. Colour is an effect of light and if colours are added this implies that different lights are added. The resultant effect on the brain is a new colour, lighter than the originals.

For example, if two monochromatic lights of wavelength 650 nm (red) and 540 nm (green) are mixed and shone onto a white surface, the eye cannot separate the colour of the two sources, but can only recognise the resultant colour mix as yellow. Plate 2 illustrates the resultant colour effect of mixing three coloured lights: red, green and blue.

The red, green and blue can be called the *primaries* and the resulting yellow, cyan and magenta the *secondaries*. Any three colours can be used as primary sources provided none of them can be obtained by mixing the other two sources, e.g. no mixtures of green and blue will produce red.

Yellow is sometimes referred to as a complementary colour to blue, or as 'minus blue', since the resultant mixture will then contain all three primaries. Similarly, cyan ('minus red') is complementary to red.

The additive mixing of colour is used in many ways, for example:

- Light sources. The mixing of coloured lights is the basis of stage lighting. Primary colours are used and by suitable dimming control, any desired colour effect can be obtained.
- Modern lamps. The mixing of colours takes place within the lamp enclosure. For example, the high-pressure mercury lamp in its uncorrected state produces a cold white light, deficient in red. This is partly remedied by introducing a phosphor, yttrium vanadate, which emits red light.
- Colour television. There are three colour transmitters, each activating a separate pattern of dots on the screen. Enlarged through a magnifier a screen would appear as a mosaic of red, green and blue dots. At normal size the eye cannot resolve the dots but only their resultant colour mix. A breakdown of the green transmitter would cut out cyan, green and yellow, leaving a darker picture in shades of blue, magenta and red.

Subtractive

Most materials which reflect light are selective in what they absorb. A black-body radiator would absorb all incident light and appear black at room temperature. Dark red paint absorbs much of the blue and green, reflecting only red. It has subtracted the blue and green from the light. Similarly, a red filter transmits only the red, subtracting the blue and green. Adding two paints or filters has a cumulative effect, and this is illustrated in Plate 3. Whenever subtractive mixing occurs the resultant colour is darker than the original.

An appreciation of the subtractive effect is essential when relating the colour performance of lamps to the colour of the lamp shades and room surfaces. A deep red decor will appear orange if the lamps used are deficient in red. This cannot be overcome by using a red-tinted shade or diffuser around the lamp as this will only reduce the general level of illumination without improving the colour.

Self-assessment task 3.2

A blue object is illuminated by a lamp that is deficient in red and blue light. What colour will the object appear and will the use of a blue filter be of any help?

Specification of colour appearance of surfaces

The colour of surfaces is a combination of the spectral reflecting properties of the surface and the spectral composition of the light source. It may be safest to judge and select colours under filament lamps or daylight as it is quite likely that certain colours will be slightly or even seriously distorted under certain discharge lamp lighting.

The nearest to a universal method of colour specification is the Munsell system.

The Munsell Colour Atlas

Munsell was an American artist and teacher and in 1915 he devised a systematic classification of colour. The Munsell Colour Atlas applies only to coloured surfaces and it is understood and used by many architects and interior designers. Colour is specified in three terms:

1. *Hue* is a description of the actual colour, e.g. red, green.
2. *Value* is a measure of the 'whiteness' of the colour (0 is pure black; 10 is pure white).
3. *Chroma* is a measure of the purity of the colour or 'colourfulness'. Starting with a neutral base, and adding a coloured pigment, the chroma steadily increases until the colour fully saturates the base. Chroma can provide a degree of emphasis, and by choosing colours of different chroma it is possible to vary the emphasis in the pattern of colours. High levels of chroma and value are inevitably associated with yellowish hues, and the lower values and chromas are associated with deep reds and blues. To make full use of the complete range of colours, extremes in value and chroma should be avoided other than for specific emphasis.

BS 5252: 1976 Framework for colour coordination for building purposes

There have been a number of BS colour systems, their aim being to try to bring some sanity and unity to the use of colour in buildings. The hope is that the manufacturer can provide a limited consistent range of colour finishes, which the specifier will specify. The benefit to the user is that colours match, despite materials differing, and that the user's green acrylic bath bears some relation to the user's green china wash basin, provided both colours are to the same BS number.

BS 5252 is developed from Munsell and colours have three parameters:

1. hue – designated by an even number;
2. greyness – designated by a letter;
3. weight – designated by a further number.

There are 13 items which have a similar role to *hue* in the Munsell system:

00 = neutral
02 = red-purple
04 = red
06 = warm orange
08 = cool orange
10 = yellow
12 = green-yellow
14 = green
16 = blue-green
18 = blue
20 = purple-blue
22 = violet
24 = purple

The *greyness* represents the clarity:

A = grey
B = near grey
C = distinct hue
D = nearly clear
E = clear, vivid colour

The *weight* refers to the lightness and relates to both value and chroma in the Munsell system. This will be an odd number and the higher numbers are darker.

Individual colours are identified by a combination of a hue number, a greyness group letter and a weight number, in that order. For example, 12 B 29 means that the hue number is 12, a greenish yellow. The greyness group is B – close to grey, but with a slight hue. The weight number is 29, indicating a darkish tone. The colour is a dark yellow-green. The Munsell equivalent is: 2.5GY 2/2.

Self-assessment task 3.3

A pale blue ceramic tile is specified in Munsell terms as B6/4. What would be the approximate equivalent in terms of BS 5252?

Using the CIE diagram (Plate 1) what would be the approximate x, y chromaticity coordinates?

BS 4800 refers specifically to paint colours and a useful reference table is in the CIBSE Code.

Specification of colour appearance of light sources

Correlated colour temperature

The colour appearance of a 'near white' source can be indicated by its match with a black-body radiator and quoted by the absolute temperature of that radiator.

Table 3.1 Correlated colour temperature classes and colour rendering groups

Correlated colour temperature (CCT) (K)	CCT class
CCT ≤ 3300	Warm
3300 < CCT ≤ 5300	Intermediate
5300 < CCT	Cold

Table 3.2 Typical lamp colour properties

Type of lamp	CCT	Class	CR group
Filament GLS	2800	Warm	1A
Fluorescent tube			
Warm white	2800	Warm	3
White	3500	Intermediate	3
Triphosphor	2700	Warm	1B
	3500	Intermediate	1B
High-pressure lamps			
MBF		Intermediate	3
HPS		Warm	4
HPS deluxe		Warm	2

The range is shown in Table 3.1. Examples of lamps and their CCT are given in Table 3.2.

The most appropriate colour appearance is a matter of personal choice. The only accepted guidelines are that at low lighting levels cool colours tend to give a gloomy appearance and lamps of different colour appearance should not be mixed haphazardly. They can be mixed intentionally as part of the design, e.g. tungsten spotlights can be used for highlights in a space otherwise lit with intermediate colour fluorescent tubes.

The CIE chromaticity diagram (1931)

This is a very precise method of specifying surface colours and light source colours. Its principal use in lighting is for the British Standard specification of fluorescent lamp colours (BS 1853: 1967); it is also used in the pigment and dye industry, and to specify colours of filters.

To understand the system it is necessary first to consider the 'three primaries' method of colorimetry (the measurement of colour). Using three suitable primaries, any other colour effect can be obtained when they are mixed. J. Guild used three monochromatic primaries of blue (435.8 nm), green (546.1 nm) and red (700.0 nm). Figure 3.1 shows how, by mixing these primaries, he could match all the spectral colours. For example, a blue-green of 500 nm wavelength would be matched by 0.54 units of blue and 0.97 units of green minus 0.52 units of red, or

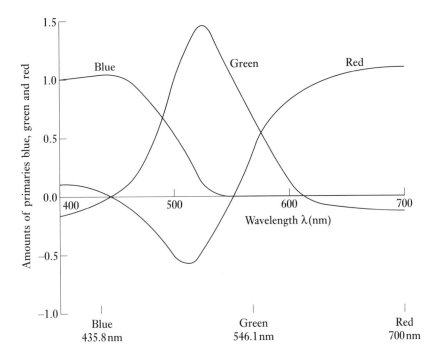

Fig. 3.1   The three primaries used by J. Guild

$$C_{500\,nm} = 0.54B + 0.97G - 0.52R$$

This introduces the concept of negative colours. By using real primaries it is impossible to match a monochromatic colour by mixing other colours, since the result will be a lighter colour. Using Munsell terms, it will match in hue but not in chroma.

This method of mixing can be represented graphically using a colour triangle, as shown in Fig. 3.2. RBG is an equilateral triangle whose corners represent the three primaries. The centre of the triangle, W, represents equi-energy white, since it is a colour made up of equal amounts of each primary.

This could be written as

$$W = \frac{1}{3}B + \frac{1}{3}G + \frac{1}{3}R \qquad\qquad [3.1]$$

Numerically $\frac{1}{3}$ = PW = QW = RW and the proportion of a primary is represented by the perpendicular distance of the point from the opposite side of the triangle to the primary.

A colour at point P would be written as

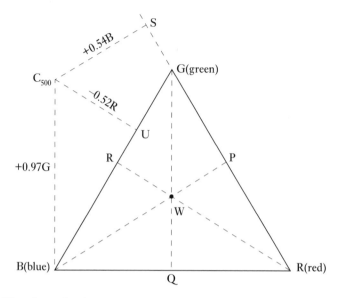

Fig. 3.2   The colour triangle

Colour P = zero B + $\frac{1}{2}$G + $\frac{1}{2}$R

The coordinates are (0, $\frac{1}{2}$, $\frac{1}{2}$).

A negative quantity will lie outside the triangle and C_{500} will be shown as in Fig. 3.2. The amount of B is represented by CS, G by CB and R by CU, where

CS : CB : −CU = 0.54 : 0.97 : −0.52

All the spectral colours (i.e. colours of a single wavelength), with the exception of 435.8 nm (B), 546.1 nm (G) and 700.0 nm (R), will lie outside the triangle. These, when plotted and joined, form the spectral curve or locus.

To overcome this 'negative' coefficient, the CIE system employs another larger triangle which encloses the spectral curve produced by the first triangle using real primaries. This is a mathematical development involving the use of 'unreal' primaries X, Y and Z. Plate 1 shows the CIE triangle.

This triangle is used in similar fashion. The X and Y coordinates x, y are measured off the axes and are called the chromaticity coordinates, where $x + y + z = 1$. It is normal to quote only the x and y values, since $z = 1 - (x + y)$.

All the spectral colours lie on the spectral locus, and all colour mixes lie within the locus.

The black-body locus is a curve representing the colour appearance at various colour temperatures of a full radiator (black body).

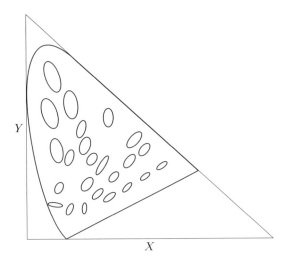

Fig. 3.3 The MacAdam ellipses on the CIE diagram

There must normally be some tolerance in colour specification and BS 1853: 1960 (amended 1962) quotes this in terms of MacAdam ellipses, which represent a maximum permissible area of deviation of the colour. These are illustrated in Fig. 3.3. This is of particular significance in the measurement of light sources.

The Y stimulus is directly proportional to the luminance of the sample colour. This means that if the y coordinate for all the spectral colours was plotted, the curve would have the same shape as the eye luminous efficiency curve. The only difference is that at 555.5 nm, $y = \frac{1}{3}$, whereas $V_\lambda = 1$.

Self-assessment task 3.4

Lamp A has CIE coordinates $x = 0.526$, $y = 0.418$, and lamp B has $x = 0.380$, $y = 0.380$. Describe their colour appearances and see if you can identify either type of lamp.

The uniformity chromaticity scale (UCS) 1960

The permissible tolerances in the specification of a colour can be indicated by an ellipse (see previous paragraph).

This has not been entirely satisfactory as an ellipse indicates an area of acceptable variation from a specified colour. Figure 3.3 shows that the ellipse varies in size depending on the part of the diagram and colour.

There have been many attempts to modify the diagram to achieve more even ellipses, and Fig. 3.4 shows the diagram replotted on a u–v axis, where

$$u = \frac{4x}{-2x + 12y + 3} \qquad v = \frac{6y}{-2x + 12y + 3} \qquad [3.2]$$

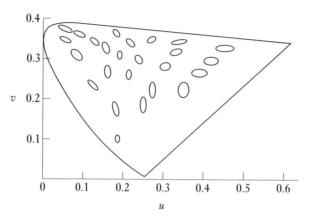

Fig. 3.4   The MacAdam ellipses on the UCS diagram

Example

A colour is red, specification $x = 0.60$, $y = 0.40$. What are the u, v coordinates?

Solution

Using eqn [3.2],

$$u = \frac{4 \times 0.6}{-2 \times 0.6 + 12 \times 0.40 + 3}$$

$$= 0.36$$

$$v = \frac{6 \times 0.40}{-2 \times 0.6 + 12 \times 0.40 + 3}$$

$$= 0.36$$

Colour rendering of light sources

The ability of a lamp to reveal colours is a basic property of its spectrum. Figure 3.5 shows a typical spectrum from three different types of lamps.

In the figure, ① is a low-pressure (LP) sodium lamp used mainly for street lighting. The spectrum is line and monochromatic (single wavelength) and the lamp will only reveal that colour. It has no ability to discriminate colours, which is the basis of colour rendering.

② is a filament lamp spectrum. It is continuous, covering all wavelengths fairly evenly with an emphasis on the red end. The colour rendering will be good.

③ is an MBF mercury discharge lamp. Most discharge lamps lie somewhere between ① and ②, and in this case the spectrum is a mixture of some lines, some continuous background spectrum due to the heat of the lamp and a band of energy in the red end created by the colour-correcting phosphor.

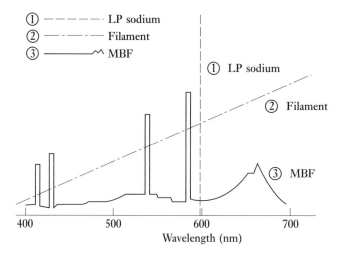

Fig. 3.5   The spectra of various lamps

Table 3.3   CIE colour rendering index

Colour rendering groups	CIE general colour rendering index (R_a)	Typical application
1A	$90 \leq R_a$	Wherever accurate colour matching is required, e.g. colour printing inspection
1B	$80 \leq R_a < 90$	Wherever accurate colour judgements are necessary and/or good colour rendering is required for reasons of appearance, e.g. shops and other commercial premises
2	$60 \leq R_a < 80$	Wherever moderate colour rendering is required
3	$40 \leq R_a < 60$	Wherever colour rendering is of little significance but marked distortion of colour is unacceptable
4	$20 \leq R_a < 40$	Wherever colour rendering is of no importance at all and marked distortion of colour is acceptable

The CIE colour rendering index (Table 3.3)

Although there are a number of different colour indices, this is by far the most widely used. The method uses eight Munsell test colours and the general index R_a is based on:

1. the spectral reflectance of the test colour
2. the spectrum of the source under test
3. the spectrum of the reference source

Although this is the main system used it has its critics and limitations. The main problem is that a single index cannot fully describe the lamp. Two lamps with the same R_a may have widely different effects on colours and some additional information would be helpful. Sometimes a deliberate distortion of colour occurs, such as with the fluorescent tube 'Deluxe Natural'. It has an enhanced red content which is favoured in the display of cooked and fresh food, but owing to the distorted spectrum it only has a relatively low R_a.

Self-assessment task 3.5

Refer to Fig. 3.5. An underpass is lit with a combination of lamps ① and ③. What colour will a red, yellow and white car appear to be?

Colour constancy

Despite all attempts to specify colour and produce standard sources, the human response to colour is variable and not always predictable.

One example is the ability to recognise that surfaces have the same spectral reflectance, even though their colour appearance is different. This is known as 'colour constancy' and an extreme example could be the lighting of a wall by different coloured spotlights. Provided it is realised that multi-coloured spotlamps are used, no assumption is made that the wall decoration is multi-coloured. There are, however, conditions when an observer cannot be sure if colour difference is due to the source or surface.

Colour matching

If colours are to be matched then it is essential that the process is carried out under controlled conditions. Only too frequently articles that appear to match under fluorescent lighting do not when taken into the daylight. This is called *metamerism* and a non-metameric match is one which matches under any lighting condition. This is only likely to happen if using identical pigments and material.

BS 950 specifies requirements for industrial colour matching, including the performance of the lamp, the construction of the viewing booth and the illuminance (at least 1000 lx).

EXERCISES

3.1   Distinguish clearly between colour appearance of a light source and its colour rendering properties.

Describe an experiment which could demonstrate that two light sources, having the same colour appearance, could have very different colour rendering properties.

3.2   Describe the Munsell system for specifying colours of surfaces.

You are asked to specify the lighting for an art gallery. In what terms could you reliably describe the colour performance of the lamp and what type would you suggest? Sketch the probable spectral power distribution of the lamp.

3.3   Two light sources have the same luminance and their CIE coordinates are $x = 0.23$, $y = 0.27$ and $x = 0.30$, $y = 0.27$ respectively. If they both have the same light output and they are put together what will be the coordinates of the resultant light?

3.4   An architect wishes to use strong colours for the room decorations. The architect's choice is deep blue Munsell B5/6 for the walls and Y6/4 for the ceiling. The proposed lighting system is uplighting, i.e. no direct lighting on the floor.

If 25 per cent of the light falls on the walls and 75 per cent on the ceiling, calculate approximately the improvement in lighting efficiency if lighter colours of B7/4 and Y8/2 were used.

CHAPTER 4

LAMPS

TOPICS COVERED

PRODUCTION OF LIGHT
Filament lamps
Tungsten halogen lamps
Low voltage display lamps
Discharge lamps
Types of discharge lamps and typical control circuits
Low pressure sodium vapour lamps
Technical data for lamps
Lamp survival
Summary of lamps and their performance

The whole of interior lighting design revolves around the lamp, and development and competition within the industry have resulted in an enormous range of lamps now being available to lighting designers. The proliferation of lamp types is a mixed blessing because lamp development can often make an existing installation obsolete both in appearance and efficiency before its time. However, a surfeit of choice is not to be scorned. It means that the lighting designer must appreciate the range of lamps available and keep up to date on new developments.

PRODUCTION OF LIGHT

The source of electromagnetic radiation has been discussed in Chapter 1. The emission of radiant energy only occurs if the electron also receives energy. The two principal methods are by heating (resulting in thermal radiation) and by collision (passing an electric current through a gas or vapour). Coupled to the collision process is fluorescence, where a phosphor is bombarded by photons as opposed to electrons.

Self-assessment task 4.1

Which element of the atom is the source of electromagnetic radiation? Is the main radiation from mercury at low pressure visible?

Filament lamps

Temperature radiation

In a tungsten-filament lamp the passage of electricity through the filament raises the temperature of the molecules within the filament to the point where they begin to give off light, or they *incandesce*. This can be explained as follows.

In any body, the molecules which make up its construction are (except at −273°C, absolute zero) in constant motion. However, the physical state of the material (solid) results in these molecules being formed into fixed patterns or structures. Consequently, their only freedom for movement is vibration about a fixed point. In general terms this vibration process is extremely complex, but the net result is that energy is given to each molecule by interaction and collision. With each vibration some of this imparted energy is released by the molecule in the form of radiation. The effect is analogous to *excitation* (see 'Discharge lamps') but in molecular form.

The energy that is radiated obeys 'black-body radiation' laws, and the resulting spectral energy distribution is a function of the temperature of the filament, as shown in Fig. 4.1. The laws are based on a physical concept of a 'black body' or full radiator. This is a body that absorbs all radiation falling upon it. The tungsten filament when heated performs in a similar if not identical pattern to a full radiator.

The spread of the spectral energy distribution is given by Planck's equation:

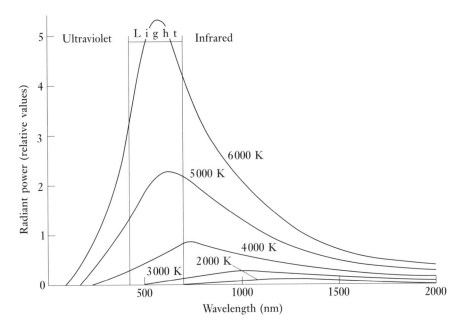

Fig. 4.1 The radiant power from a black-body radiator at different temperatures

$$P_e = \frac{C_1}{\lambda^5[\exp(C_2/\lambda T) - 1]} \qquad [4.1]$$

where P_e is the power radiated (W/m^3), λ is the wavelength (m), T is the absolute temperature of radiator (K), $C_1 = 3.7415 \times 10^{-16}$ W m^2 and $C_2 = 1.4388 \times 10^{-2}$ m K. In this expression it is seen that the power radiated per unit surface area is dependent only on its temperature.

Two further laws that affect the performance are:

1. Wien's radiation law:

$$P_{e\lambda_{max}} \propto T^5 \qquad [4.2]$$

2. Wien's displacement law:

$$\lambda_{P_{max}} \times T = \text{const} \qquad [4.3]$$

Summarising, [4.1] gives the spectral distribution at temperature T; [4.2] shows that as T increases, the maximum power increases by the fifth power; and [4.3] shows that as T increases the wavelength of peak radiation decreases. These laws combine to indicate that:

1. The vast majority of the radiant power is in the infrared.
2. Increasing the temperature not only increases the power output per unit area but also increases the proportion in the visible spectrum.

Filament lamps are efficient radiators of power but very inefficient radiators of light.

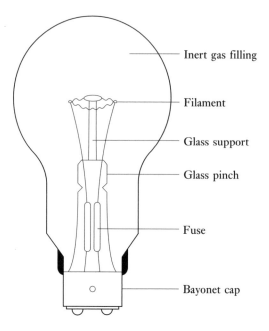

Inert gas filling

Filament

Glass support

Glass pinch

Fuse

Bayonet cap

Fig. 4.2   General lighting service (GLS) – filament lamp

Example

If the CCT of the sun is 6000 K, and a filament lamp is 3000 K, assuming the peak radiation from sunlight is at the point where $V_\lambda = 1$, at what wavelength will the peak for the lamp occur?

Solution

V_λ is 1 at 555 nm (see Chapter 1). Therefore, using eqn [4.3],

$$555 \times 6000 = \lambda \times 3000$$
$$\therefore \quad \lambda = 1110 \text{ nm}$$

The general lighting service (GLS) lamp

A lamp designed for general lighting service (GLS) is illustrated in Fig. 4.2.

Filaments   These are constructed of drawn tungsten wire which is then coiled, and in the lower-wattage gas-filled lamps the coils can be coiled again to reduce convection losses (coiled-coil).

Although the melting point of tungsten is 3600 K, at temperatures above 2800 K the rate of evaporation increases to an extent where lamp life is drastically reduced. More efficient lamps are available, e.g. in projection equipment, but they have much shorter lives.

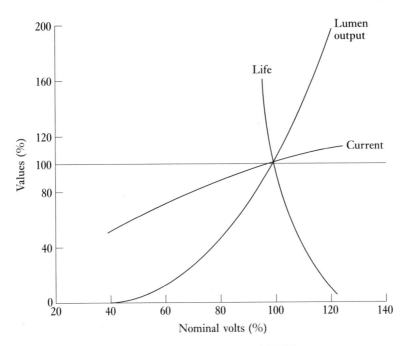

Fig. 4.3   Effect of voltage variation on the performance of GLS lamps

Gas filling   The rate of filament evaporation can be reduced by raising the vapour pressure in the lamp. This can only be done by introducing gases which are chemically inactive with hot tungsten. An added complication is the loss of heat due to convection and conduction by the filler gas. The most common gas filling is a mixture of argon and nitrogen, and this can only be used with coiled filaments where the heat losses are much less than for a straight-wire filament.

Glass envelopes   These are spherical or 'mushroom' in shape and can be clear, pearl (etched on the inside surface), or 'inside white' (where the inside is coated with silica, titania or some similar substance). Clear and pearl bulbs have the same efficacy. 'Inside white' bulbs have 4–8 per cent lower efficacy but a much greater degree of diffusion.

Lamp caps   In the UK GLS lamps normally have bayonet caps (BC) up to 150 W. Edison screw (ES) caps are used for 200 W lamps and Goliath Edison screw (GES) caps are used for 300–1500 W lamps.

Lamp life   Figure 4.3 shows that small variations in supply voltage cause dramatic changes in life. There are instances where lamps are deliberately overrun to improve their performance. This is true of projector lamps which may only have a life of a few hours in order to achieve much greater temperature and efficacy.

Conversely, where lamps are difficult to maintain they may be deliberately underrun to extend their life.

Example

A 100 W GLS lamp operating on 240 V has a light output of 1260 lm and life of 1000 h. What will be its performance on 260 V?

Solution

Referring to Fig. 4.3, a 20 V increase represents $(20/240) \times 100 = 8.3$ per cent. Values can only be obtained approximately from the graph, but 8 per cent will produce a 40 per cent life and 140 per cent light change. Therefore,

$$\text{Life} = 0.4 \times 1000$$
$$= 400 \text{ h}$$
$$\text{Light output} = 1.4 \times 1260$$
$$= 1764 \text{ lm}$$

Tungsten–halogen lamps

The most important factor deciding the light output of a lamp is the filament temperature, and this determines the life of the lamp.

The addition of a halogen vapour to the gas filling effectively reduces the evaporation rate of the filament. This allows the use of a higher filament temperature and, hence, promotes greater efficacy without the detriment of short lamp life.

The tungsten–halogen regenerative cycle

The halogens are a group of elements comprising fluorine, chlorine, bromine and iodine which are chemically very reactive with other elements. With the addition of a halogen to the gas filling, a complicated chemical cycle is established whereby the evaporated tungsten particles combine with the halogen to form a reusable compound called a *halide*.

This compound will condense at temperatures below 250°C but by constructing the lamp in such a way that the bulb wall temperature is maintained above this value, the tungsten halide is prevented from condensing and will be swept around the inside of the lamp by convection currents. Eventually it will pass near the filament region where the high temperature will cause the compound to dissociate into tungsten, which is deposited back onto the filament, and halogen, which is released to repeat the cycle. Tungsten–halogen lamps therefore do not blacken, and full light output is maintained throughout life.

Construction

The bulb must be capable of operating continuously above 250°C and a form of fused silica called quartz is used. To maintain lamp wall temperatures above 250°C

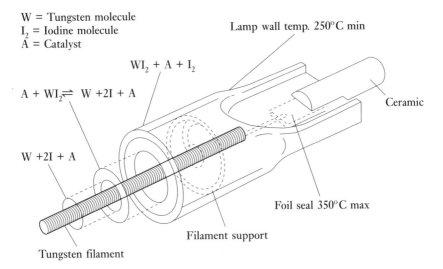

W = Tungsten molecule
I_2 = Iodine molecule
A = Catalyst

Lamp wall temp. 250°C min

$WI_2 + A + I_2$

$A + WI_2 \rightleftharpoons W + 2I + A$

Ceramic

$W + 2I + A$

Foil seal 350°C max

Filament support

Tungsten filament

Fig. 4.4 Tungsten–halogen linear lamp

requires a small bulb size which, in itself, has the added bonus of allowing the use of expensive but efficient krypton or other rare gases as a filling gas and in allowing the gas-filling pressure to be raised without risk of explosion. Consequently, these lamps operate with higher filament temperatures, resulting in increased efficacy and an increased lamp life, generally 2000 h. A typical linear lamp is illustrated in Fig. 4.4.

Operation

There are several operating conditions which must be observed for satisfactory performance:

- The quartz bulb wall must be maintained above the condensation temperature of the halide (approx. 250°C).
- The hermetic seal between the quartz bulb and the molybdenum lead-in wire must be kept below 350°C. Above this temperature the molybdenum lead-in wire starts to oxidise and a mechanical stress is applied to the constituents of the seal, which breaks.
- The coolest part of the filament (where it enters the lamp) must be maintained above a critical temperature. Should it not do so, then some corrosion of the filament wire may take place with a reduction of life. For this reason it is not advisable to reduce the voltage to the lamp to less than 90 per cent of its rated value, except where this parameter has been considered by the lamp manufacturer during design.

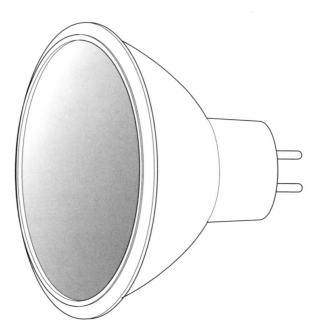

Fig. 4.5 Tungsten–halogen lamp (narrow spot) with dichroic reflector (courtesy Sylvania Lighting Products Ltd)

- Linear lamps must be operated within 4° of horizontal. Outside this range, the halogen vapour will migrate to the lower end of the lamp, resulting in early lamp failure of the upper part of the filament.
- Contamination of the outside surface of the lamp wall must be avoided. For example, a deposit of grease by handling will cause the quartz envelope to fail at high temperatures as the grease will cause the surface to develop fine cracks (devitrification). Before use, any such contamination should be cleaned off with a suitable solvent such as methylated spirit.

Low-voltage display lamps

There is a wide range of lamps available for use in optical systems or with their own integral reflector. This can be an internal aluminium coating or a dichroic coating (see Chapter 5). Using the advantages of the halogen cycle they are very compact and have a lamp life often in excess of 2000 h. A typical lamp is shown in Fig. 4.5 and it has a diameter of 50 mm or less.

The reason for the improved optical performance is the much thicker and shorter filament required to operate at 12 V. A transformer is necessary, but the lamps and components are safer to handle. They may be used as single units, or as a group fixed to a low–voltage trunking system.

Self-assessment task 4.2

What is the function of nitrogen in a GLS lamp?

If a 230 V GLS lamp is run at 215 V what effect will this have on the life of the lamp?

Why is quartz used in the halogen lamp?

Discharge lamps

The method of light production previously discussed is concerned with the passage of electricity through a solid. Lamps which produce light owing to the passage of electricity through a gas are termed *discharge* lamps.

In the early 1900s the theoretical work of Max Planck led Niels Bohr to propose a model of the atom, similar in form to our own solar system. The atom consists of a central, dense, positively charged nucleus surrounded by one or more relatively light, negatively charged particles or electrons. These electrons are attracted towards the nucleus but maintain their distance by moving at high speed in fixed orbits around it, in much the same way as our planet orbits the sun.

The electrons moving in the innermost orbits are those most strongly bound to the nucleus; the electrons in outer orbits are more loosely held and are more likely to be involved in the process of collision and subsequent radiation.

Ionisation

Initially when a voltage is applied to a discharge lamp the gas inside the lamp is in an insulating state and will not conduct electricity.

To make the gas conductive a process of ionisation must occur. Normally each gas atom is electrically neutral, i.e. the number of orbiting electrons balances or equals the number of positive charges within the nucleus. By the application of energy to the lamp, outer electrons can be freed so that they become independent negatively charged particles, or *negative ions*. Such a process also leaves the parent atom with a deficiency of one negative charge, so it becomes a *positive ion*. This ionisation process must occur continuously inside the lamp to enable conduction to take place and to replace ions that are lost by recombination.

Ionisation can occur within the lamp in two ways. Initially the application of a high-voltage pulse to the lamp will pull many electrons from their orbits. However, once the lamp is running, charged particles moving along the lamp will collide with gas atoms dislodging electrons in the process.

Excitation and light production

The charged particles move along the lamp because they are attracted towards the electrodes at the end of the lamp.

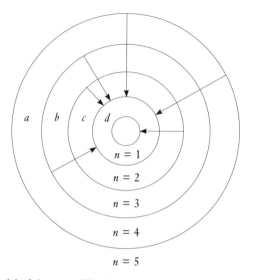

Fig. 4.6 Bohr's model of the atom. The circular orbits represent stable energy levels for a single electron. It cannot exist in the areas *between* orbits *a*, *b*, *c*, etc.

As they move towards it they accelerate. Many collide with gas atoms before they have gained sufficient energy (speed) to cause ionisation. In these cases the outer electrons are not totally dislodged, but are displaced to a higher orbit.

Bohr showed that displacement can only occur to specific orbit radii and that electrons cannot exist in areas *a*, *b*, *c* or *d* shown in Fig. 4.6.

This displacement is termed *excitation*. An electron may be excited from, say, orbit 2 to orbit 3 or 4 and then return to orbit 2 either directly or from 4 to 3 to 2. At each transition the energy difference between each orbit will be released in the form of photons of radiant power.

High- and low-pressure discharges

In the low-pressure discharge the charged particles move along the lamp a relatively long distance between collisions and will easily build up sufficient speed to cause excitation to the first possible orbit. As the gas pressure is increased the distance between collisions decreases and so the energy gained by the particle during its acceleration time is proportionately lower. However, a study of a practical high-pressure lamp and the power radiated from the discharge shows that the power is due to excitation to higher possible orbits. This may appear to conflict with the basic atomic concept.

The answer is to be found in the collisions which occur before the accelerating particle has had time to build up sufficient speed to cause excitation. In the low-pressure discharge their number is relatively low and has no great effect on the discharge. When the pressure is raised 3000–4000 times, the energy released by each of these collisions produces sufficient heat to raise the temperature of the gas. This high temperature (6000°C) enables many outer electrons to possess sufficient energy to move to the first possible orbit unaided. Thus the collision processes cause excitation from this point.

Additionally, the increased number of gas atoms present tends to trap any radiation from this first orbit by reabsorbing it and then releasing it many thousands of times as it travels towards the surface of the lamp. The method of producing radiation is basically the same at all pressures, the differences being that at high pressure the atoms combine to form molecules and the energy emissions are at the higher orbit levels.

It is interesting to note that as the first-orbit radiations are reabsorbed, the important resonant radiation at low pressure is completely absent at high pressure.

Fluorescence

The low-pressure mercury discharge produces radiant energy within the ultraviolet (UV) part of the spectrum. By coating the inside of the arc tube with a fluorescent coating or phosphor, this UV radiation is converted to visible radiation.

The phosphor must have a strong absorption in the short-wave UV band around 253.7 and 185 nm. It must have very low absorption in the visible range and it must have optimum performance at 40–50°C. Table 4.1 shows a typical range of phosphors and their emission. The colour is also affected by metal activators used with a phosphor.

High-pressure mercury lamps also make use of phosphors, but in this case the excitation energy is at 365 nm and the phosphors are required to produce light only in the red region. Phosphors used include magnesium fluoro-germanate, and with the improved deluxe lamps, yttrium vanadate.

Table 4.1   Typical fluorescent lamp phosphors

Phosphor	Colour of light
Calcium halo-phosphates	White
Magnesium fluoro-germanate	Red
Calcium tungstate	Blue

Gas and vapour fillings in discharge lamps

The discharge envelope is filled with a mixture of gases and vapours. The main gas or vapour is the one responsible for the emission of light. If a vapour, this may be in solid state at room temperature, so a further gas is needed to initiate the discharge. When the lamp is running there may be considerable loss of energy due to the current electrons not colliding with the main gas or vapour, and even passing out of the envelope. To increase the probability of collision a buffer gas or vapour is included.

Examples of these three elements are given in Table 4.2.

Table 4.2 Typical gas and vapour fillings

Lamp type	Main filling	Starting	Buffer
Fluorescent lamp	Mercury	Argon	Argon
Low-pressure sodium	Sodium	Neon	—
High-pressure sodium	Sodium	Xenon	Mercury
High-pressure mercury	Mercury	Argon and nitrogen	—

Discharge lamp control gear

During starting it is necessary to introduce a higher than normal voltage to the lamp to assist ionisation. However, once the gas has begun to conduct, its resistance will progressively fall (as more and more gas atoms are ionised), until, if unchecked, the mains current will become excessive and the circuit fuse will fail.

To prevent this occurring it is necessary to impose some control on the current that can flow through the circuit. This control equipment or gear could take the form of a large ohmic resistor, but this is wasteful, dissipating considerable amounts of power as heat. A more satisfactory current-limiting device is an inductive resistor, termed a *choke* or *ballast*, which is essentially a coil of wire wrapped around a metal core. Current flowing through the wire produces a magnetic field which then hampers the growth of further current thereby 'choking' the current to the desired level.

This type of control only operates on an alternating current supply and in the process creates an overall circuit lagging power factor, where

$$\text{Circuit watts} = \text{volts} \times \text{amps} \times \text{power factor} \qquad [4.4]$$

The watts are unaffected, but a low power factor results in an increase in the supply current. In the UK the power factor is normally corrected from 0.5 (uncorrected) to 0.85. It is not normal to correct for lamps below 30 W rating.

All discharge lamps require control gear of some sort from the humblest of neon indicators to the largest stadium floodlight. The size, weight and cost of such equipment is in proportion to both the lamp wattage and the complexity of lamp technology being used.

Example

If a 58 W fluorescent lamp operates on 240 V, 50 Hz supply and the lamp voltage is 100 V, what is the choke inductance necessary to control the lamp circuit? To simplify the calculation assume the choke has no resistance, i.e. is a 'pure' inductor.

Solution

The circuit will be as shown in Fig. 4.7. If the lamp power is 58 W and it is assumed to act as a resistor, then

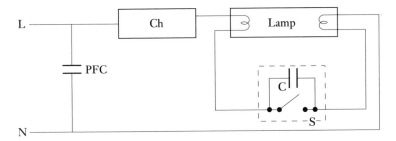

Fig. 4.7 Switch-start circuit

Power = V × A
58 = 100 × A

∴ A = $\dfrac{58}{100}$

= 0.58 amps

This is the circuit current (I_C).

The choke is a pure inductance and its voltage (V_{Ch}) is 90° out of phase with the lamp voltage (V_L). Therefore,

$$V_{mains} = \sqrt{V_L^2 + V_{Ch}^2}$$

$$240 = \sqrt{100^2 + V_{Ch}^2}$$

∴ $V_{Ch} = 218$ V

The impedance of the choke is V_{Ch}/I_C. Therefore,

$$\frac{V_{Ch}}{I_C} = 2\pi \times \text{frequency} \times L$$

where L is the inductance in henrys. Thus,

$$\frac{218}{0.58} = 2\pi \cdot 50 \cdot L$$

$$L = 1.20 \text{ H}$$

Types of discharge lamps and typical control circuits

Tubular fluorescent lamp

This is basically a low-pressure mercury discharge lamp, producing a high proportion of its radiation in the UV region at 253.7 and 185 nm. This is converted to the visible region by the fluorescence of the phosphor coated on the inside of the glass tube. Table 4.3 summarises the energy change for a 'white' fluorescent tube.

Lamp construction and performance   The low-pressure mercury vapour lamp can have a relatively high light output and efficacy provided the UV radiation

Table 4.3 Energy conversion of 58 W fluorescent lamp

Lamp type	Conducted and convected heat	Short-wave UV radiation			Visible light
Plain 58 W lamp, no phosphor	35%	62%			3%
Lamp with phosphor added	35%	16%	30% Infrared	16%	3%
Total		51%	30%	19%	

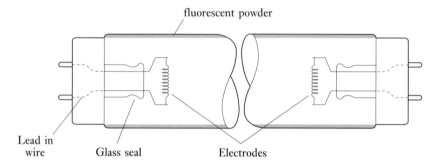

Fig. 4.8 Tubular fluorescent lamp (MCF)

generated in the lamp is converted into light by means of the fluorescent coating. This type of lamp is one of the most common lamps manufactured today. It is used extensively in industrial and commercial premises and is beginning to be used in the home. Its general construction is shown in Fig. 4.8.

Electrodes The electrodes, which are at the ends of the tube, act as both anode and cathode when connected to alternating current but are normally called cathodes.

Electrons are emitted at the cathode and collected at the anode. For an electron to leave the cathode, it has to be given sufficient energy to overcome the forces holding it to the atoms of the cathode material. This energy is called the 'work function' and different materials have different work functions. The cathode is therefore coated with a thin layer of emissive materials, such as the alkaline earth oxides of barium, strontium and calcium, which have a low work function and emit electrons freely on heating.

The main cathode material used is tungsten (because of its low evaporation rate) in the form of a coiled-coil or triple-coil filament or a braided filament, which is coated with the emissive material.

Heating the electrodes by passing a current through them before the arc is struck makes sure that a large number of free electrons exist around the electrodes, making the striking of the lamp easier. During operation the electrodes are kept hot by the passage of the discharge current.

Fluorescent coating (phosphor)   The amount of light depends upon the combination of phosphors used. The wavelengths radiated by the phosphor vary with the chemicals used. For example, the phosphors used in present general-purpose fluorescent lamps are *halo-phosphates* which contain calcium, antimony, chlorine, fluorine and manganese. With no manganese the colour of the lamp is blue and, by adding different amounts of manganese, blue-white and white to yellow-white lamps are obtained. A range of 'white' lamps is produced from 'colour matching' to 'warm white', each having a different colour appearance and colour rendering properties.

One important effect of the choice of phosphors for a given colour appearance is that, in general, the better the colour rendering, the lower the light output. Since the introduction of new-generation 'hexagonal aluminate' phosphors it has been possible to provide both high-efficiency light output and deluxe colour rendering with a variety of colour appearances. The phosphors are called 'narrow band' producing emissions in the 455–485 nm band (blue), 525–560 nm band (green), and 595–620 nm band (red). These lamps may be called 'triphosphor'.

Control gear   In fluorescent lamp circuits operated on a.c. supplies, the ballast is usually an inductance (choke) and is connected in series with the lamp, as shown in Fig. 4.7. There are many more complex circuits used to control the lamp current and to assist in striking the lamp, and the term *ballast* is used to cover all forms of control gear.

Switch-start circuits

Glow-type starter switch   This switch, shown in Fig. 4.9, has two switch contacts, one of which is a bimetal strip, in a small glass envelope containing argon.

When the supply is switched on the bimetal strip is cool and the contacts are open. Current flows through the choke, through the first electrode, through the argon in the starter switch (in the form of an arc), through the second electrode and back to the supply. The arc in the argon heats the bimetal strip causing it to bend towards the other contact of the starter until the two contacts close. Current still flows through the two electrodes and the starter switch contacts, and the electrodes are pre-heated, liberating electrons into the arc tube. The argon arc no longer operates and as the source of heat is removed the bimetal strip cools and begins to straighten until the contacts open again.

The opening of the switch causes the lamp arc to be struck because a voltage pulse is induced in the choke and because of the pre-heating of the electrodes. When the arc is struck the voltage across the lamp falls to about its stabilised value.

As the starter switch is connected across the lamp, the voltage across the two contacts is also reduced but is not enough to cause an arc in the starter switch argon gas, and so they remain open until the lamp is required to be lit again. If the lamp does not start the starter operates again until the lamp eventually strikes or the supply is switched off.

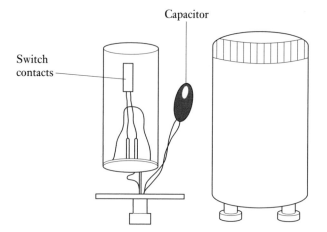

Capacitor

Switch
contacts

Fig. 4.9   Glow-type starter switch

Starter capacitor   The small capacitor which is connected across the starter
terminals (i.e. in parallel with the switch contacts) reduces emission by the lamp at
radio frequencies and also improves the waveform of the starting voltage pulse.

Circuit operation on starting   Having considered the sequence of events that
takes place in the starter switch, this must be related to the lamp and choke during
this time. First there is a current flowing through the choke, cathodes and starter
switch. This current is about 1.5 times the running current; it heats the cathodes
and causes free electrons to be emitted and ionisation to take place near the
electrodes. When the starter switch opens, the choke opposes the change by
producing a voltage which is in the same direction as the supply voltage and
therefore adds to it. A high-voltage surge across the lamp is normally enough
to cause an arc between the electrodes of the lamp.

Electronic starter switch   A switch is available, at extra cost, which uses a
thyristor trigger. This sets up a pre-heat electrode circuit and a high voltage is
generated. Starting is near instant and, in the event of lamp failure, the electrode
current ceases. This prevents damage to the starter and ballast.

Starterless circuits

Although the switch-start circuit is reliable and is more suitable for low-
temperature applications, the starter switch is the one item in the circuit with
moving parts liable to fail. Also, in some applications its operation can be a nuisance
owing to flicker until the lamp has struck – especially with lamps towards the end
of their life.

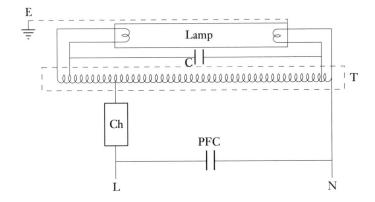

Fig. 4.10 Starterless transformer circuit

Starterless circuits operate without a starter switch and so reduce the maintenance required. They have disadvantages, however: they are more complex and expensive; ballast losses are higher than those of the simple choke; and they are more susceptible to difficulties in starting at temperatures below 5°C. A circuit is shown in Fig. 4.10.

The major change from the switch–start circuit is that a transformer is used to provide a low voltage across each electrode. This is contained within the ballast canister together with the choke.

When the circuit is first switched on, nearly all the supply voltage appears across the lamp and the transformer. Current flows through the electrodes from the two transformer windings, one heating each electrode and producing emission of electrons, until there are enough free electrons to enable the lamp to strike with only the supply voltage across the lamp. Once the arc has struck the current through the choke increases and therefore the voltage drop previously across the transformer now appears across the choke, reducing the lamp and transformer voltage so that the lamp current is limited to its correct value. The voltage across the electrodes falls to about half.

The circuit is suitable for use with dimming control. To dim fluorescent tubes the lamp voltage and filament temperatures must be maintained while the lamp current is reduced. A typical circuit is shown in Fig. 4.11.

Semi-resonant start circuit

This comes within the category of starterless circuits, but is simpler and has lower gear losses than the transformer type. It comprises a double-wound choke with two equal coils wound in opposition and a series capacitor. The term *resonant* is used since the subcircuit of the lamp, coil B and the capacitor (Fig. 4.12) produce a high lamp voltage and filament current for starting (the capacitor and coil combination have minimum impedance). When the lamp is running, coil B becomes ineffective

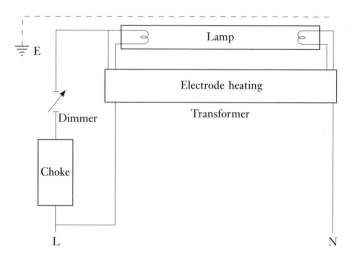

Fig. 4.11   Circuit for dimming a fluorescent tube

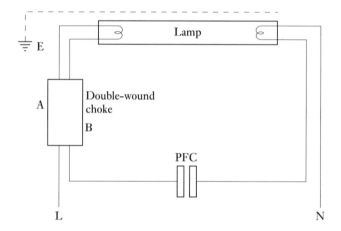

Fig. 4.12   Semi-resonant start circuit

and the circuit is coil A, the lamp and the capacitor. The circuit will operate at low temperatures. Although this circuit may not be in production now, it is in use in many existing luminaires.

Electronic control circuits

Raising the electrical frequency to the lamp from 50 Hz to a high frequency of around 30 kHz brings a number of significant improvements to performance. These include:

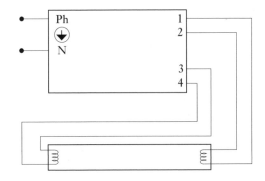

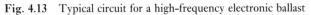

Fig. 4.13 Typical circuit for a high-frequency electronic ballast

- reduction in ballast losses and weight
- elimination of flicker and noise
- improved lamp efficacy and life
- instant lamp start

The electronic element of the circuit is the frequency conversion. Figure 4.13 shows an example of this type of circuit.

The sequence of events is:

1. Convert the incoming mains 50 Hz to direct current (rectify).
2. By means of an oscillator reconvert the direct current to alternating current at 30–100 kHz frequency.
3. Utilising an inductive circuit of greatly reduced size and weight, start and control the lamp.

The total consumption of the high-frequency (HF) lamp and ballast is considerably less than that of conventional fluorescent lighting, typically saving 30 per cent of energy cost for equivalent light output. Even more savings are available for air-conditioned buildings, where a smaller-capacity plant could be installed owing to the reduced heat load from the HF lighting.

The HF system offers particular advantages in twin-lamp luminaires, because a single ballast drives two lamps instead of the two separate ballasts in conventional luminaires.

Table 4.4 compares HF energy effectiveness against conventional switch-start lamps in a widely used luminaire type. This highlights the halving of control gear losses, and the big jump in overall circuit efficacy.

Self-assessment task 4.3

Give two reasons why using HF control will raise the circuit lumen efficacy.

Table 4.4   Comparison of HF with convention circuits

	Conventional luminaire with 38 mm MCF lamps	HF lamps and luminaires 26 mm lamps
Lamp wattage	2 × 65 W	2 × 50 W
Control gear losses	2 × 12 W = 24 W	1 × 12 W = 12 W
Total consumption	154 W	112 W
	(taken as 100%)	(saving 29%)
Light output	2 × 4990 = 9980 lumens	2 × 5000 = 10 000 lumens
Overall circuit efficacy	65 lm/W	89 lm/W

Lamps for starterless circuits

The starterless circuit does not produce a high-voltage pulse to strike the lamp, and so lamps have been developed which can start, after a short period of pre-heating, on the normal mains voltage. These lamps may also be used on switch-start circuits.

There are two types designed for the starterless circuits. One (MCFA) has a thin metal strip running the length of the tube and connected at each end to the metal lamp caps which must be earthed. The other (MCFE) has a silicone coating over the outside of the lamp. It has been found that the voltage required to start a lamp varies with the lamp surface resistance and that the voltage required is at a minimum if the resistance is *very high* or *very low*. The metal strip provides a very low resistance, the silicone coating a very high resistance. A further advantage of an earthed strip is that usually the neutral of the supply is connected to earth and therefore the strip is at approximately the same potential as the neutral. As the strip is close to the electrodes (separated by the glass) a high-voltage flow exists between the strip and the electrode which is connected to the live terminal. This high potential assists in ionisation.

To assist the silicone lamp, an earthed metal surface at least 25 mm wide and less than 10 mm away from the lamp, running the length of the lamp, is required. The metal spine of the luminaire normally satisfies this requirement.

Effect of changes in ambient temperature on lamp output and starting

If the air temperature around the lamp is high, the pressure of the mercury vapour inside the lamp rises and the light output of the lamp decreases. If, on the other hand, the air temperature is low, the vapour pressure will not be high enough for sufficient collisions to take place between the free electrons and the mercury atoms, and the light output goes down, as shown in Fig. 4.14.

There is a narrow band of temperature at which the light output is a maximum. Normal fluorescent lamps are designed to run in an air temperature of about 25°C, with a lamp wall temperature of 40°C which produces a vapour pressure of about 800 Pa. When lamps are operated above or below this temperature the light output will be less than the declared value. Lamps are available for use in an ambient temperature of 35°C.

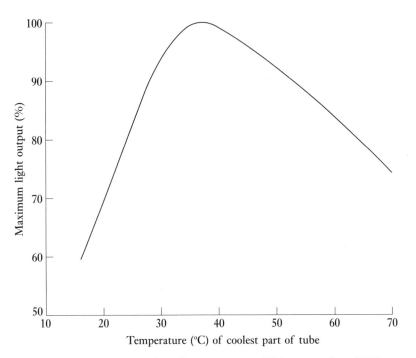

Fig. 4.14 The relation between tube wall temperature and light output for a 65 W MCF lamp

The vapour pressure inside the lamp can be reduced by simply cooling a single area of the lamp, because the excess mercury will condense at the cool spot. For example, heat can be conducted away from the lamp by placing a metal shoe in contact with the outer surface of the lamp. The shoe has to be held in position by a device holding it to the luminaire body (through a hole) and it must be spring loaded to maintain adequate contact. Various other methods have been used, such as setting the electrodes well into the lamp or fixing an insulating shield between the electrodes and the lamp cap to keep the ends of the lamp cool, but all these methods have been superseded by the *amalgam* (TLH) lamp.

In the amalgam lamp, cadmium or indium is inserted in the lamp either on the electrode mount or in the form of a separate ring on the inside of the glass envelope. The indium combines with the mercury in the lamp, forming an amalgam of mercury and indium. The amalgam stabilises the mercury vapour pressure at high temperatures and maintains a high light output.

Self-assessment task 4.4

Which discharge lamps are temperature sensitive and what steps can be taken to reduce the effect of overheating?

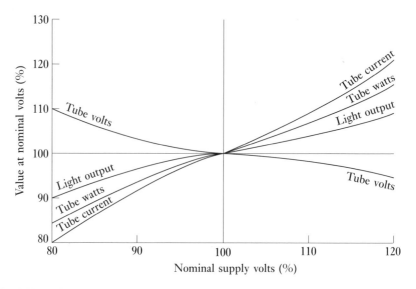

Fig. 4.15   Effect of voltage variation on the performance of a 40 W MCF lamp

Effect of changes in supply voltage

The changes in volts, light output and lamp current that occur with changes in supply voltage are shown (for the switch-start circuit) in Fig. 4.15.

The changes are not as drastic as those produced by variation of the voltage of a tungsten lamp, e.g. a 1 per cent change in voltage produces only a 1 per cent change in light output.

Effect of change in supply frequency

The normal changes of ±1 per cent in supply frequency have about the same effect on inductive circuits as a 1 per cent voltage change. Operating the lamp on a normal supply other than 50 Hz, say 60 Hz, has no effect on the lamp characteristics provided the control gear is designed for the frequency used. Operating the lamp at frequencies above 400 Hz improves the light output, and reduces the fluctuations in light output, the size of the control gear, and the losses. These improvements are quite significant, and designing control gear to operate at 400 Hz and above is worthwhile when such a supply is available – e.g. in aircraft, or for operating from low-voltage supplies such as in caravans and boats. When connecting to a normal 50 Hz system the cost of changing the frequency can outweigh the advantages gained.

Compact fluorescent lamps

The majority of new lamps have been designed for general lighting of interiors and exteriors. There is a significant section of lighting, concerned with corridors, toilets, desk lights and the home, where the GLS lamp has been the major choice.

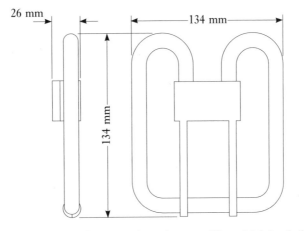

Fig. 4.16 A 2D 16 W compact fluorescent lamp (courtesy Thorn Lighting Ltd)

Table 4.5 Comparison of GLS lamps with a range of compact fluorescent lamps

GLS (W)	Initial lumens	2D (W)	Initial lumens	PL (W)	Initial lumens
60	610	16	1050	9	400
100	1230	28	2050	11	600
150	2060			15	900
				20	1200

This market is now challenged by the low-wattage fluorescent tube, bent into shapes that enable it to fit into small luminaires. Figures 4.16 and 4.17 show two such lamps. The PL lamp can be used as direct replacement to a GLS lamp. The 'D' lamp is used in luminaires specifically designed for it. Table 4.5 compares their performance with GLS lamps. However, one word of caution: power factor correction is not normally required for lamps below 30 W and the saving in volt amperes may not be as satisfactory as the apparent saving in watts.

QL induction lamps

This is a form of low-pressure fluorescent lamp which does not require any form of electrodes within the discharge tube. The lamp's current is induced by a magnetic ferrite core with a coil wound around it as shown in Fig. 4.18. The induced current sets up collisions as already described previously, with the generation of UV radiation. The lamp is controlled by an HF electronic circuit and the result is a life in excess of 60 000 h and an efficacy of 60–70 lm/W.

The main drawback is that the lamp is relatively expensive. Its applications are in situations where lamp access is difficult and maintenance is expensive.

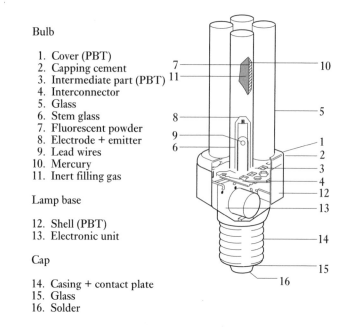

Bulb

1. Cover (PBT)
2. Capping cement
3. Intermediate part (PBT)
4. Interconnector
5. Glass
6. Stem glass
7. Fluorescent powder
8. Electrode + emitter
9. Lead wires
10. Mercury
11. Inert filling gas

Lamp base

12. Shell (PBT)
13. Electronic unit

Cap

14. Casing + contact plate
15. Glass
16. Solder

Fig. 4.17   PL compact fluorescent lamp (courtesy Philips Lighting Ltd)

Low-pressure sodium vapour lamp

Light is produced by the action of the excited electrons returning to lower-energy levels. The sodium vapour arc produces a number of spectral lines but most of the visible radiation (95 per cent) is concentrated in a very narrow region of the spectrum at 589.0 and 589.6 nm (known as the sodium 'D' lines). The light output is therefore *monochromatic* (single coloured) and the lamp light is easily recognised by its yellow appearance.

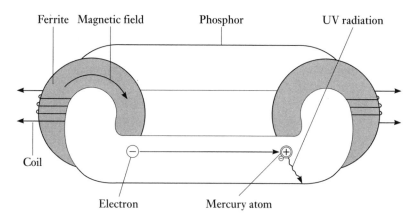

Fig. 4.18   Osram version of induction lamp: two ferrite coils wrapped around the tube induce in the gaseous filling a current of electrons which causes mercury atoms to emit ultraviolet radiation which in turn excite the coating of the glass, causing it to phosphoresce

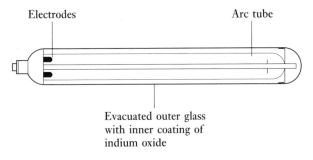

Fig. 4.19 Single-ended low-pressure sodium lamp (SOX)

Although this has the disadvantage that colours are greatly distorted, it has the advantage that, as the relative response of the eye (V_λ) is high at 589 nm (approximately 0.76), it has a potential for a high lumen output, depending upon how much energy is actually put into the lamp.

It is now possible to manufacture low-pressure sodium lamps with efficacies of about 200 lm/W, but not all of the lamps produced are this efficient and the efficacy varies with the lamp wattage.

Lamp construction and performance

The main type of lamp is SOX, which is a single-ended lamp with an integral outer jacket coated with a heated reflector (see Fig. 4.19). The inner discharge tube is made of two-layer glass, the outer being soft glass and the inner a thin film of sodium-resistant glass.

The electrical characteristics are similar to those of the fluorescent tube. The lamps are temperature sensitive but they operate within a vacuum jacket.

Lamp circuits

At room temperature the metallic sodium is solid, and in order to vaporise it, neon and argon are added. Initially a neon–argon arc is formed and the heat produced gradually vaporises the sodium. The lamp has a run-up period of about 10 min before maximum light output is reached.

A transformer is used in the sodium lamp circuit because of the 'higher-than-mains' ignition voltage required, i.e. approximately 470–490 V. If a normal transformer is used, a choke would also be needed to limit the current after the discharge has struck because of the negative resistance characteristic of the arc. Such a circuit is, however, expensive in material usage, when one considers that the transformer output voltage is about 500 V while the normal operating lamp voltage required is 100 V.

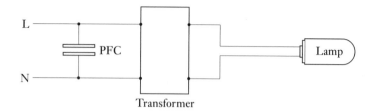

Fig. 4.20   'Leakage-flux' transformer circuit

Material can be saved by, first of all, using an *auto*-transformer and, secondly, by making it an *auto-leak*-transformer which removes the need for a separate choke (see Fig. 4.20).

This leak transformer type of circuit has a low power factor (0.3) and a capacitor is connected across the supply to increase it to an acceptable value.

High-pressure mercury lamp (MBF)

Increasing the vapour pressure of a discharge has the effect of changing the spectral power distribution by increasing the width of the spectral lines (bands) produced, and by changing the relative intensities. Absorption of the resonance radiation also takes place.

Lamp construction and performance

The basic high-pressure mercury vapour lamp, shown in Fig. 4.21, consists of a glass outer envelope containing the quartz arc tube, its supports, and the lead–in wires. This outer envelope contains nitrogen to dissipate the heat produced and maintain the arc tube at the correct temperature.

Inside the arc tube are a small quantity of mercury and argon (which acts as a starting gas), and the tungsten electrodes. There are usually three or four electrodes – two of which are main electrodes and a smaller auxiliary electrode (or electrodes) used to initiate the arc. The auxiliary electrode is very close to a main electrode and is connected through a resistor to the electrode at the other end of the lamp. At switch-on the full mains voltage exists between the auxiliary electrode and its adjacent main electrode. To assist in electron emission, the main electrodes are impregnated with an emissive material of a low work function.

Lamp operation

At room temperature the mercury in the lamp is liquid and the argon is included in the arc tube to start the arc and to evaporate the mercury. When the lamp is connected to the supply, there will initially be no current flow and mains voltage will exist between the two main electrodes and between a main electrode and its

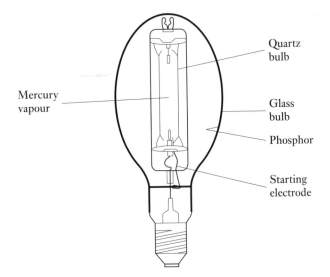

Fig. 4.21   High-pressure mercury fluorescent lamp (MBF)

auxiliary electrode via the internal resistor. This voltage will produce a local arc in the argon gas with the current limited to a safe value by a high resistance of a few thousand ohms connected in series with the auxiliary electrode as shown in Fig. 4.21. The initial arc produces ionisation, and an argon arc is soon struck between the two main electrodes. This produces further ionisation until sufficient mercury ions exist for the mercury arc to be formed. At first a low-pressure arc exists and very little light is produced; but gradually, as the lamp heats up, the mercury vapour pressure rises and a high-pressure arc is formed and more light is emitted. Once the main arc is established, its resistance is lower than in the auxiliary electrode circuit and this auxiliary electrode ceases to function.

The time taken for the lamp to 'run up' completely – i.e. to reach full light output – is approximately 5 min. Should the mains voltage be reduced to approximately 80 per cent or less of the normal value the lamp will be extinguished, and as the pressure is high it will not reignite unless a very high voltage is used or until the lamp has cooled and the pressure reduced. Although the subsequent run-up period is reduced, the total time to reach full light output again from switching off is of the order of 10 min.

MBF lamps have a fluorescent coating such as magnesium fluoro-germanate or yttrium vanadate on the inside of the outer envelope to convert the long-wave UV radiation into visible energy, especially in the region of the spectrum above 600 nm, thus improving the spectral power distribution and, hence, the colour rendering.

Owing to the fairly good colour rendering, relatively high efficacy, and wide range of sizes available, this type of lamp is now extensively used in outdoor lighting and in industrial high-bay installations. MBF lamps are also mixed with filament lamps in display lighting.

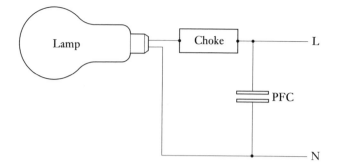

Fig. 4.22 Control circuit for an MBF lamp

This type of lamp is also used for street lighting in residential areas where its colour rendering may be preferred to low-pressure sodium.

Lamp circuits

The circuit is relatively simple as the lamp incorporates its own starter. The circuit is shown in Fig. 4.22 and consists of a choke and power factor capacitor.

A series resistance in the form of a tungsten filament can be used instead of the choke. These are called MBT- or ML-type lamps; they have improved colour rendering but only about half the luminous efficacy. MBF lamps, in common with all high-pressure lamps, cannot at present be satisfactorily dimmed.

Self-assessment task 4.5

Give an example of a situation where the MBF lamp may be preferable to the fluorescent lamp, giving the reasons why.

Metal halide lamps (MBI)

Known as the mercury halide or metal halide lamp, this lamp is similar in construction to the MB lamp but has a much shorter arc tube. Strictly speaking, it is not a mercury lamp but has a mixture of elements combined within a single arc tube. When such metals as dysprosium, thallium, indium or sodium are introduced into the mercury arc tube, the mercury spectrum is suppressed (reducing the UV radiation), but the added metal atoms partly compensate by causing a wide range of spectral bands to be emitted. It is possible to tailor the spectral distribution to produce either a 'white' light or, alternatively, a predominantly coloured light (e.g. green), depending upon the relative proportions of the metals used. These metals have the disadvantage that, used in their elemental form, they would attack the discharge tube. Therefore, halides of the metal are used, the most common being the *iodide*, e.g. sodium iodide.

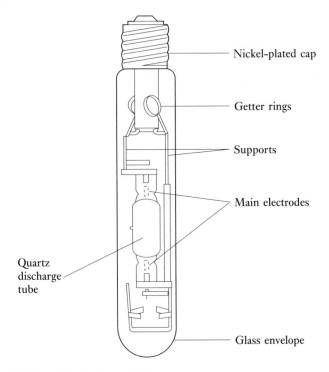

Nickel-plated cap

Getter rings

Supports

Main electrodes

Quartz
discharge
tube

Glass envelope

Fig. 4.23　375 W metal halide lamp (MBI)

One disadvantage of these lamps is that a high starting voltage is necessary. This is usually provided by an electronic ignitor which generates very short-duration high-voltage pulses and is connected across the lamp terminals. This type of starter also enables the lamp to be restruck within 1 minute of being switched off.

Owing to its high efficacy and good colour rendering properties, the MBI lamp is suited to both high-bay interior and high-mast exterior installations, and is often used in sports stadiums, especially where colour television transmission is expected. The small arc tube enables good optical control of the source light output and thus the lamp is suitable for use in floodlighting projectors. A typical lamp is shown in Fig. 4.23. A version is also available with a phosphor coating – MBIF.

Lamp circuits

The lamp requires an ignitor, choke and power factor capacitor. A typical circuit is shown in Fig. 4.24.

High-intensity discharge (HID) lamps

This is an extension of the metal halide range operating at higher pressures with a corresponding reduction in size and increase in arc luminance. The combination

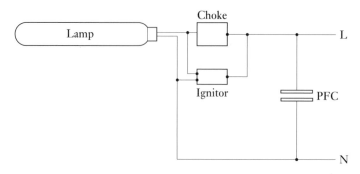

Fig. 4.24   An example of a control circuit with ignitor for an MBI lamp

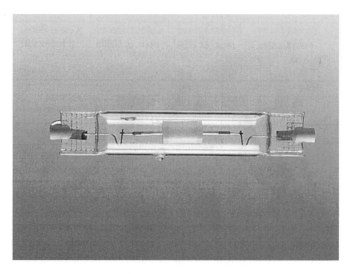

Fig. 4.25   Philips Mastercolour HID lamp

of small size and a colour rendering index (R_a) above 80 makes the lamp ideal for floodlighting and projection systems. Figure 4.25 shows a 150 W lamp which is 135 mm in overall length and has an efficacy of 90 lm/W.

High-pressure sodium lamp (SON, SON-L)

The low-pressure sodium arc emits all of its radiation in the two yellow lines at 589 and 589.6 nm. The colour rendering is, as stated earlier, unsuitable for most applications other than road lighting and security lighting.

If the arc vapour pressure is increased beyond the optimum pressure, increased absorption of the 'D' lines occurs and the light output is reduced. However, as the pressure is increased and the 'D' lines disappear, other broad bands of energy appear in the spectrum until an almost continuous visible spectrum is emitted.

This type of source, because it produces radiant energy at the extremes of the visible spectrum where the V_λ value is low, has a lower efficacy of approximately 100 lm/W but provides much better colour rendering than the low-pressure lamp.

Note that the efficacy, despite the reduction by comparison with low-pressure sodium, is higher than can be obtained from mercury and metal halide sources.

Lamp construction and performance

The high temperature (1300°C) necessary to produce the high vapour pressure required in this lamp would melt any glass arc tube, and at this temperature quartz also is readily attacked by hot sodium vapour. It was not possible, therefore, to construct a practical SON high-pressure sodium vapour lamp until the early 1960s when *alumina*, a 'sintered' aluminium oxide with a small percentage of magnesium oxide, was first used.

Alumina can withstand the sodium attack. It is a translucent crystalline material and transmits about 90 per cent of the light produced by the arc. It is not possible to pinch-seal the electrodes in the arc tube as can be done with glass or quartz, and special niobium caps containing the electrodes have been developed which are sealed to the arc tube ends by a sophisticated form of brazing.

A 400 W high-pressure SON sodium lamp is shown in Fig. 4.26.

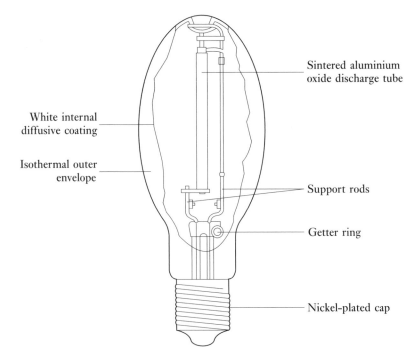

Fig. 4.26 High-pressure sodium lamp

The lamp consists of an evacuated outer envelope containing the arc tube and its support – similar to the MB lamp except that the arc tube is longer and narrower and no auxiliary electrode is possible. The arc tube contains an amalgam of mercury and sodium together with a small quantity of argon or xenon to assist in striking the arc. The mercury is used to reduce the lamp current for a given wattage.

A high voltage is required across the lamp in order to strike the arc. This high voltage may be provided, as in the MBI circuit, by an electronic ignitor which generates pulses of a few thousand volts for a few microseconds' duration and is connected across the lamp terminals. The ignitor continues to pulse until the lamp is struck.

Some manufacturers include in the outer envelope of the lamp a bimetal starting switch and small heating element. The switch and element are connected across the electrodes.

This type of lamp is manufactured in numerous sizes from 50 W (70 lm/W) to 400 W (120 lm/W) and the outer envelope may be either elliptical (E) with a diffusing coating on the inside, or tubular (T) clear glass.

The lamp sizes are similar to the high-pressure mercury vapour lamps and direct replacement may be made by using sodium lamps, providing the control gear is suitable. Some lamps are designed to be a direct replacement for a particular mercury vapour lamp without any major modifications to the circuit, e.g. a 330 W plug-in SON sodium lamp may replace a 400 W mercury lamp, giving approximately 25 per cent more light for 15 per cent less power (370 circuit watts against 420 watts). These replacement lamps, however, are not as efficient as the normal SON lamps and, therefore, are not recommended for new installations. The high-pressure sodium SON lamp is an economic light source and is used where colour rendering is not too critical. Typical applications are street lighting, outdoor area (car parks etc.) and building floodlighting; also industrial lighting, especially in high-bay-type installations. The increase in the cost of energy is causing an increasing number of indoor applications to be successfully and acceptably converted to high-pressure sodium sources.

Self-assessment task 4.6

Why is the internal coating on the SON-E lamp not a phosphor, as used for the coating of the MBF and MBIF lamps?

SON deluxe lamps

Further increase in the sodium vapour pressure causes a reduction in luminous efficacy but significant improvement in colour rendering. The deluxe lamps achieve an improvement in CRI from group 4 to group 2 and can be used in most interiors where a warm colour appearance is acceptable. The luminous efficacy is around 70 lm/W for the higher-wattage lamps.

A range of low-wattage SON-white lamps are available for use in display lighting. They have an efficacy of 40–50 lm/W and a group 2 colour rendering.

Lamp circuits

As with other discharge lamps the circuit comprises a starter, choke and power factor capacitor. The starting process was outlined in the previous section.

Technical data for lamps

In order to select the best lamp for a specific application and carry out lighting calculations, technical data are published in many lamp catologues.

Apart from dimensions, weights and electrical performance the following data are useful.

- Spectral distribution. Plate 4 shows the spectral distribution of most of the lamps discussed in this chapter. It shows the relative power radiated at different wavelengths.
- Lumen maintenance. Figure 4.27 will show the percentage change in lumen output during the lamp life.

Figure 4.27 compares a single-phosphor white fluorescent lamp with a triphosphor lamp. The difference is of great significance when determining a maintenance factor – see Chapter 7.

Self-assessment task 4.7

What type of lamp has 100 per cent lumen maintenance?

Lamp survival

It is often difficult to determine the life of lamps as they do not all fail after a set period. A number of types, like old soldiers, simply fade away! As with lumen maintenance, this information is needed for determining realistic maintenance factors. An example of a survival curve is shown in Fig. 4.28.

As well as this information, it is useful to know the colour rendering (CRI) and colour appearance (CCT), and whether or not the lamp light output is dependent on ambient temperature.

Summary of lamps and their performance

No single table can show the complete range of lamps, but Table 4.6 indicates the main groups of lamps, their range of availability and their performance. *Detailed information is only available from the manufacturers' catalogues.*

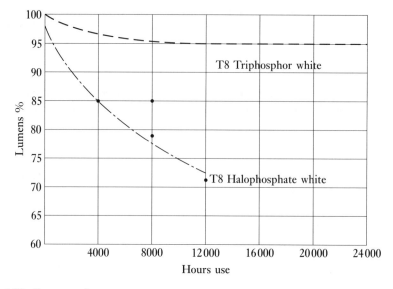

Fig. 4.27   Lumen maintenance curves

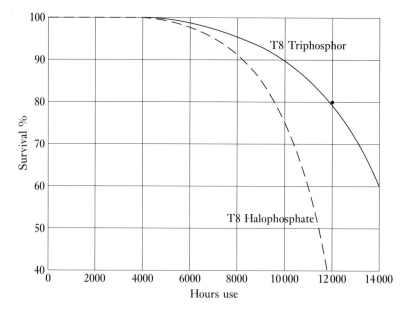

Fig. 4.28   Lamp survival curves

Table 4.6 Summary of lamp types and their characteristics

Characteristics	Tungsten filament		Low-pressure discharge		High-pressure discharge				
	GLS	TH	MCF	SOX	SON	SON deluxe	MBF	MBI	QL
Wattage range (W)	25/2000	50/2000	4/125	18/135	50/400	50/400	50/1000	70/2000	75
Efficacy range (lm/W)[a]	8/18	18/24	37/100	100/200	70/130	70/85	35/58	65/85	65
Colour appearance (CCT class)	Warm	Warm	Warm/ intermediate /cold	Yellow	Warm/ yellow	Warm	Cold	Intermediate/ cold	
Colour rendering group	1A	1A	1A/3	—	4	2	3	2 or better	
Need for ballast	No	No	Yes	Yes	Yes	Yes	Yes	Yes	Yes
Starter/ignitor	No	No	Yes[b]	No	Yes[c]	Yes	No	Yes	No
Available with internal reflector	Yes	Yes	Yes	No	Yes	No	No	No	No
Typical use	Home	Display	Office and shop	Streets	Industry	Office (uplighting)	Industry	Industry and commerce	Difficult access
Average life (h)[d]	1000/2000	2000/4000	9000/16 000	16 000	24 000	12 000	24 000	15 000	60 000+

[a] This is lamp watts only.
[b] On switch-start circuits.
[c] Some lamps have internal switches.
[d] Life to 50% survival. (This varies with manufacturers.)

EXERCISES

4.1   Compare the effect of overrunning a GLS lamp and a fluorescent lamp by 10 per cent mains voltage. Explain the reasons for the difference.

4.2   Compare the advantages and disadvantages between standard switch-start and high-frequency control of fluorescent lamps.

Using the method on p. 74 calculate the inductance required to control a 70 W fluorescent lamp, assuming the lamp operates at 120 V and the supply is 230 V 50 Hz.

4.3   A church is lit to 100 lx with pendant luminaires each taking three 60 W GLS lamps. The proposal is to use three compact fluorescent 9 W PL lamps.

(a) What effect will this have on the illumination level?

(b) Are there any environmental factors to be considered?

(c) Is there an economic case if the lighting is only used for 500 hours per annum?

4.4   A shopping mall is to be illuminated from a height of 20 m. Access is relatively expensive. Discuss the range of suitable lamps available and make your selection, giving reasons for your choice.

CHAPTER 5

LUMINAIRES

TOPICS COVERED

CONSTRUCTION OF LUMINAIRES
   Material and finishes

OPTICAL DESIGN
   Reflectors
   Curved reflectors
   Design of louvres
   Refraction
   Diffusion

MECHANICAL CONSTRUCTION
   Luminaires in flammable atmospheres
   Photometric performance of luminaires
   Production of photometric data

LUMINAIRE APPEARANCE

The luminaire is the equipment which contains the lamp. Its four purposes can be identified as:

1. Connecting the lamp to the electricity supply.
2. Controlling the light emitted by the lamp.
3. Protecting the lamp from a hostile environment.
4. Providing a fixture of satisfactory appearance.

The relative importance of each factor depends on the use. All luminaires must satisfy 1 and be electrically safe, but many decorative designs are far more concerned with 4 than 2.

CONSTRUCTION OF LUMINAIRES

The shape of a luminaire and its method of construction are related to the lamp shape. Those for use with bulb-shaped lamps normally have a symmetrical form around the vertical axis, and their contours can be spun. This enables glass blowing and metal spinning to be used to considerable optical and decorative effect.

Tubular lamps need luminaires formed by bending, pressing or extruding into linear shapes and their designs are usually in sheet steel or plastic.

Material and finishes

Steel

This can be fabricated by bending, pressing or spinning. It is normally finished in stoved enamel which is based on the use of an alkyd resin. Where a very high standard of finish is needed an acrylic stoved enamel is generally used. Vitreous enamelling is the application of a thin layer of glass fused onto the metal surface at a temperature of about 800°C. This gives good corrosion resistance but can only be applied to shapes that will not distort at high temperatures, and do not have sharp edges where the enamel will easily chip.

Aluminium

The light weight of this metal makes it suitable for spun shapes, but it is not rigid enough for general use in fluorescent luminaires except for louvre construction. Its reflecting properties depend on its purity, the commercial grade having a reflectance around 0.7 and 0.8, and the super-purity grade having a reflectance of nearly 0.9. Although stoved enamel finish can be used, a good reflection is obtained by polishing the metal and anodising the surface to form a relatively hard oxide film, which may be transparent and which protects the reflector from corrosion. It is also used in cast form for exterior and industrial-type luminaires. The grade of alloy is normally LM6 which does not need protective finishes. Recent developments have enabled high reflectance surfaces to be applied to commercial grades of aluminium.

Plastics

These can be divided into two broad groups, acrylics and polystyrenes.

Acrylic material is supplied in sheets or granules and is available in clear, opal or coloured forms. It is shaped by pressing while softened by heat, or extruded into a continuous shape. It is a thermoplastic material.

Polystyrene is supplied in powder form and has similar properties but does not have such good stability and tends to be brittle. It is used in the cheaper ranges of diffusers and controllers.

Glass

Owing to its weight, glass is little used in fluorescent luminaires. Its resistance to abrasion, its sparkle, and the fact that it can be blown into many shapes and colours make it, in many respects, preferable to plastics in decorative tungsten lighting. Glass is obtainable in both heat-resistant and heat-absorbent forms, but at present there seems little likelihood that glass will be replaced by plastic for vertical window glazing.

OPTICAL DESIGN

Figure 5.1 shows the principal methods of optical control and Fig. 5.2 shows how these methods can be applied in the design of a typical luminaire for a fluorescent lamp.

Reflectors

Any surface which is not perfectly black will reflect light. The amount it reflects and the way in which it is reflected can be defined as the reflection property of that surface. Even a sheet of clear glass reflects some of the light incident on it (about

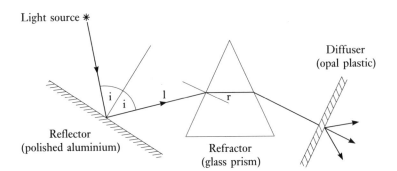

Fig. 5.1 Methods of light control

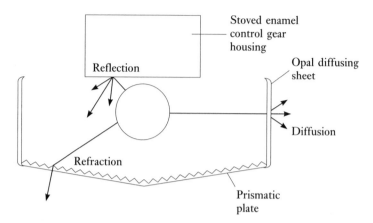

Fig. 5.2   Examples of light control in a fluorescent luminaire

8 per cent). It is usually the reflective properties of materials which reveal objects. Print is readable because the printing ink has different reflection properties from the paper on which it is printed. The light received is the same in both cases, but the light reflected differs.

Snell's laws of reflection

The two laws of specular reflection (i.e. reflections from highly polished mirror types surfaces), are:

1. The *incident ray*, the *normal* to the surface and the *reflected ray* are in the same plane.
2. The *angle of incidence* is the same as the *angle of reflectance*.

These are illustrated in Fig. 5.3 and they apply, whatever the shape of the reflector, at the point of incidence of the ray.

Specular reflectance

This is measured by the proportion of the light reflected in accordance with the laws of specular reflection. A good specular reflector, such as a silver-backed mirror, may have a specular reflectance near to 0.95, i.e. it *reflects* 95 per cent of the light it receives. Table 5.1 gives the reflectance of typical surfaces which can be used as specular reflectors.

Diffuse reflectance

If the surface is rough or non-shiny it will still reflect light (unless it is perfectly black), but this reflected light will obey the laws of specular reflection at minute

Table 5.1

Material	Finish	Reflectance
Aluminium (commercial grade)	Anodised and polished	0.7–8
Super-purity	Anodised and polished	0.9
Aluminised plastic	Specular	0.94
Stainless steel	Polished	0.60
Chromium	Plate	0.66

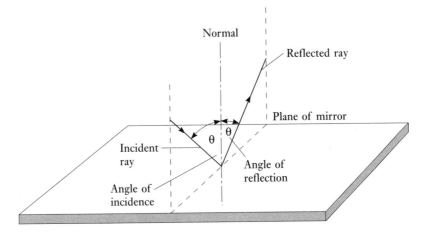

Fig. 5.3   Reflection of light on a specular surface

irregular elements of the surface which, in effect, face different directions. It will be spread in various ways, and Fig. 5.4 illustrates different types of reflection that are identified.

Diffuse reflectance occurs when the light is scattered equally in *all* directions: a 'perfectly diffusing' reflector has the same luminance irrespective of the direction in which it is viewed. A surface which is almost a perfectly diffusing reflector is a block of magnesium carbonate. A more familiar surface is a sheet of blotting paper.

Mixed reflectance

Most surfaces have mixed reflection properties, i.e. some specular reflectance and some diffuse reflectance. A typical surface is the stove enamelling on the metal spine of a fluorescent luminaire.

Self-assessment task 5.1

Describe the type of reflection from the following:

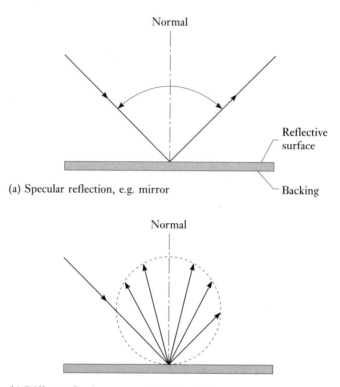

(a) Specular reflection, e.g. mirror

(b) Diffuse reflection, e.g. matt white paint

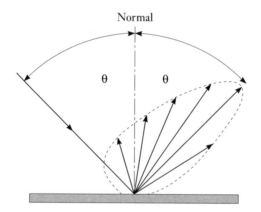

(c) Mixed reflection, e.g. stoved enamel finish

Fig. 5.4   Different types of reflection

(a) opal plastic
(b) commercial-grade aluminium
(c) plane glass

Curved reflectors

Shaping the mirror can make the reflected light rays *converge* (concentrate) or *diverge* (spread). Curved mirrors (or reflectors) are used in spotlights, floodlights and projection systems. The basic curves are *spherical, parabolic* or *elliptical.* These shapes are shown in Fig. 5.5.

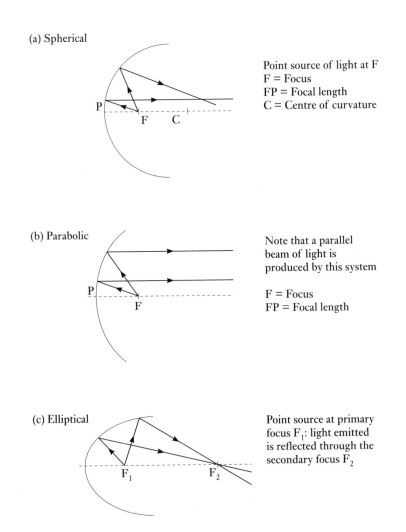

(a) Spherical

Point source of light at F
F = Focus
FP = Focal length
C = Centre of curvature

(b) Parabolic

Note that a parallel beam of light is produced by this system

F = Focus
FP = Focal length

(c) Elliptical

Point source at primary focus F_1: light emitted is reflected through the secondary focus F_2

Fig. 5.5 Curved reflector contours

Spherical reflector

If a light source is placed at the focus F, which is at half of the radius from the centre, light reflected near P will emerge very roughly parallel.

If the light source were placed at its centre of curvature, C, then reflected light will come back through C. An example is the design of the crown-silvered lamp in spotlights used in display, as shown in Fig. 5.6.

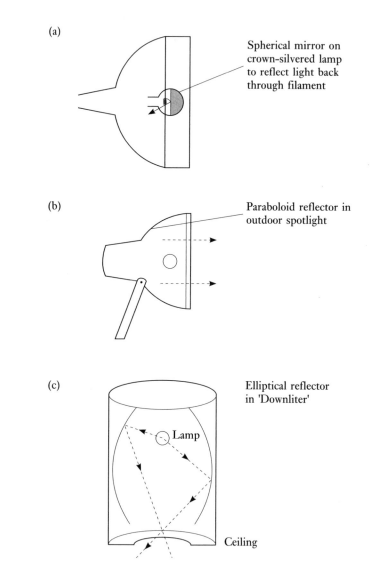

(a)   Spherical mirror on crown-silvered lamp to reflect light back through filament

(b)   Paraboloid reflector in outdoor spotlight

(c)   Elliptical reflector in 'Downliter'

Lamp

Ceiling

Fig. 5.6   Applications of reflector contours in luminaires

Parabolic reflector

The most important property of a parabola is that when a small light source is placed at its focus, all reflected light rays are parallel. If the source is moved away from the focus, the reflected rays will diverge and this will spread the beam of light. Parabolic reflectors are used for most spotlights and floodlights.

If the light source is tubular – e.g. a linear tungsten-halogen lamp – then the parabola is formed only in the plane at right angles to the lamp axis. The reflector is called a 'parabolic trough'. However, if the light source is comparatively small and of roughly equal dimensions in all directions – e.g. the filament of a GLS or projector lamp – then the reflector is parabolic in all planes, i.e. it is a 'paraboloid'.

Elliptical reflector

An ellipse has two focal points, F_1 and F_2, and if the light source is at F_1 all the light will be reflected through F_2. An ellipse can be drawn with two pins, a loop of string and a pencil. Using a plain piece of paper, draw a line down the centre and mark F_1 and F_2 100 mm apart. Tie two drawing pins so they are joined by a piece of thread 160 mm long. Fix the pins at F_1 and F_2. By holding a pencil inside the thread, stretch the thread out and hence trace out the ellipse. The major axis will be 160 mm long and the minor axis 120 mm.

The elliptical mirror is used in the design of some 'black-hole' recessed downlighters. The lamp is at the top focus and the second focus is at ceiling height. A fair proportion of the light emerges, but it will not illuminate the surrounds of the aperture which will appear dark. This is shown in Fig. 5.6(c).

Reflector size

Reflector shape controls the direction of reflected light, but the beam intensity is very dependent on reflector size or apparent area. Figure 5.7 shows two parabolic reflectors suitable for tubular sources. If the lamp is at the focus both will produce parallel beams of light but the intensity of reflector B in the forward direction (reflecting the lamp intensity) will be twice that of reflector A. If the beam is parallel, looking at the reflector, an image of the lamp will be seen emerging from all points of the surface. The reflector is said to be 'fully flashed'.

Example

If the lamp in Fig. 5.7 has a luminance in all directions of 10 000 cd/m^2 transverse to its axis, and if the specular reflectance is 0.8, calculate the intensity for A and B.

Solution

The luminance of the image in the reflector in the forward direction will be 10 000 × 0.8 cd/m^2.

The *intensity* in the forward direction will be

Luminance times the *'projected' area* of the reflector

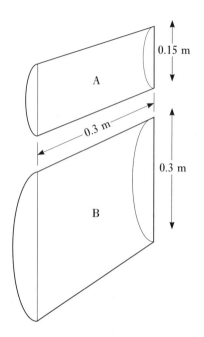

Fig. 5.7   Example of reflector size and intensity

If the angle of rectangle A is 0.3 m wide × 0.15 m high, and the area of rectangle B is 0.3 m × 0.3 m, then

$$I_A = 10\,000 \times 0.8 \times (0.3 \times 0.15) \text{ cd}$$
$$= 360 \text{ cd}$$
$$I_B = 10\,000 \times 0.8 \times (0.3 \times 0.3) \text{ cd}$$
$$= 720 \text{ cd}$$

Hence, I_B has double the intensity of I_A.

Note that the intensity of the lamp in the direction of view itself has been ignored because its contribution would be negligible in this type of reflector.

Design of louvres

A growing awareness for the need to control glare to a much larger extent than by using diffusers has led to a great increase in the use of louvres. A louvre in its simplest form is a baffle which allows light in certain specified directions but blocks it where either a direct or a reflected view of the lamp could cause glare.

Figure 5.8 shows a simple box louvre which will cut off direct views of the lamp above 30° to the downward vertical. If the louvre finish is matt black this form of 'cut-off' will be achieved but the system will be inefficient and 'dead' in appearance.

If the louvre is made of polished reflecting material then a lamp reflection could be seen at high angles of view. This is shown in Fig. 5.9.

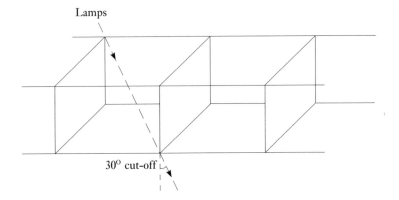

Fig. 5.8 Black box louvre

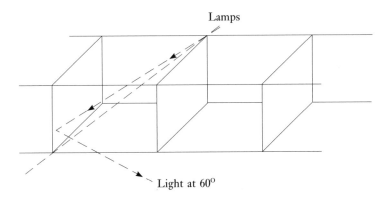

Fig. 5.9 Specular box louvre

If strict control is needed, as specified in LG3 (see Chapter 8), then the louvre blades need to be more than vertical strips. A typical shape is shown in Fig 5.10. There are problems with this type of louvre. It has to be made of high-grade reflecting material, which raises costs. The louvre will appear dark as no light should be coming out at high angles. This can give the room an oppressive feel as the ceiling will be dark. It may well be necessary to provide some form of upward or wall lighting to relieve the gloom.

The use of louvres in computer situations is discussed further in Chapter 8.

Refraction

Figure 5.11 shows the path of a ray of light through a clear glass prism. When light passes from one clear medium to another – e.g. glass to air, air to water – some light is reflected at the surface and some is transmitted. If the light enters the boundary between the two media along the normal (i.e. at right angles to the

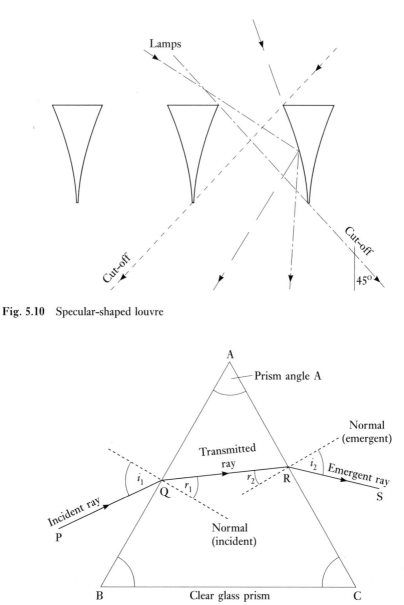

Fig. 5.10   Specular-shaped louvre

Fig. 5.11   Refraction of light through a prism

surface) its direction will not change. If it enters at an oblique angle, then its direction will change.

The angle of incidence (i) is the angle the ray of light makes with the normal when passing from one clear material to another, e.g. from air to glass, plastic or water. The angle of refraction (r) is the angle the redirected ray makes with the normal.

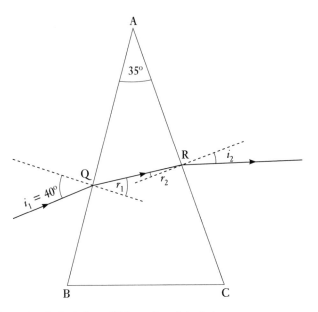

Fig. 5.12 Example of calculation of light path and deviation

When passing from a less dense material (e.g. air) to a denser material (e.g. glass) the angle of refraction (*r*) is always less than the angle of incidence (*i*):

Refractive index (μ) = $\dfrac{\sin i}{\sin r}$ [5.1]

Example

The refractive index (μ) of air to glass is 1.5. If the angle of incidence is 40°, what is the total deviation?

Solution

Figure 5.12 shows a prism and the light path can be traced. At Q the angle of incidence i_1 is 40°. Hence,

$$1.5 = \frac{\sin 40°}{\sin r}$$

$$\therefore \quad \sin r_1 = \frac{0.6428}{1.5}$$

$$r_1 = 25.37°$$

The ray emerges at R.

The geometrical calculation is as follows. Given that the two base angles ABC and BCA, of the triangle, are equal and angle BAC = 35°, each angle is (180° − 35°)/2 (since the sum of the angles of a triangle is 180°), or 72.5°. That is,

Angle ABC = angle BCA = 72.5°

In the quadrilateral BQRC, *the sum of the four internal angles* is 360°. Thus

$$\text{Angle BQR} = (90° + r_1) = 90° + 25.37° = 115.37°$$

and

$$\text{Angle QRC} = (360° - 115.37°) - (2 \times 72.5°)$$
$$= 99.63°$$
$$\therefore \quad r_2 = 99.63° - 90°$$
$$= 9.63°$$

Again

$$\frac{\sin i_2}{\sin r_2} = 1.5$$
$$\therefore \quad \sin i_2 = 1.5 \times \sin 9.63°$$
$$i_2 = 14.53°$$

Now the ray of light would have been deviated by a total angle of

$$(i_1 + i_2) - (r_1 + r_2) = (40° + 14.53°) - (25.37° + 9.63°)$$
$$= 19.53°$$

Total internal reflection occurs when the angle of incidence reaches a certain value. This means that instead of the light emerging at R it is totally reflected back into the prism.

This condition occurs when in theory the emergent angle i_2 closely approaches 90°. When this happens the glass–air barrier acts as a mirror. The angle (r) can be found since

$$\frac{\sin 90°}{\sin r} = 1.5 = \mu \quad \text{(refractive index)} \tag{5.2}$$

but

$$\sin 90° = 1$$

and hence

$$\sin r = \frac{1}{1.5} = 0.6667$$
$$\therefore \quad r = 41.8°$$

This is known as the *critical angle* for a refractive index of 1.5. All transparent materials have their own critical angle. Prisms can therefore be made to *reflect* as well as *refract*.

Diffusion

Many luminaires make no use of reflection or refraction to control the light. A domestic lamp shade may provide the right sort of light distribution in an attractive way, but it is not the result of careful optical design!

A lamp shade does alter the light distribution, partly by absorbing some of it, partly by reflecting some and partly by scattering the light that it transmits. This process of scattering is called *diffusion* of light, and materials that do this, such as opal plastic, are called *diffusers*. These should not be confused with prismatic enclosures which, if correctly designed, should be controlling and not scattering the light.

Dichroic coatings

This is a rather specialised treatment of reflectors and lenses, but it has considerable significance in the filtering out of radiant heat. A dichroic mirror consists of a glass base on which alternate layers of transparent material are vacuum deposited. Each has a different refractive index, e.g. magnesium fluoride (1.38) and zinc sulphide (2.30). If the optical thickness is $\lambda/4$, then interference will occur at that wavelength λ, and while the reflected light at each surface will be in phase, the transmitted light will be reduced to zero.

The thickness and refractive index of the layers dictate the values at which interference will occur, and so at certain values no interference will occur and most of the light will be transmitted.

A reflector or filter can be designed either to reflect or to transmit certain colours, or it can reflect or transmit light but not infrared. This provides a so-called 'cool beam', but it must be remembered that if the radiant heat is removed from the beam, it will emerge elsewhere, usually inside the luminaire.

Fibre optics

This is a specialised technique used in some forms of display and exhibition lighting. A fibre optic tube is capable of transmitting light from a source to a target some distance from the source.

A harness will comprise a number of these tubes, each able to provide a 20° beam with intensities of 200 cd or more. An example of its use would be the lighting of a display cabinet with the light source housed elsewhere; hence light with very little heat and UV and more space for the displays.

MECHANICAL CONSTRUCTION

Luminaires can take many different forms but all have to provide support, protection and electrical connection to the lamp. In addition, luminaires have to be safe during installation and operation and must be able to withstand the surrounding ambient conditions. The standard which covers most luminaires in the UK is BS 60598 (previously BS 4533) *Luminaires*. It is suitable for use with luminaires containing tungsten–filament, tubular fluorescent and other discharge lamps running on supply voltages not exceeding 1 kV. It covers the electrical, mechanical and thermal aspects of safety.

Luminaires are classified according to the degree of protection they provide against electric shock, the degree of protection against ingress of dust or moisture and according to the material of the supporting surface for which the luminaire is designed.

Table 5.2 lists the classification of luminaires according to the material of the supporting surface for which the luminaire is designed.

Table 5.2   Classification of luminaires according to the material of the supporting surface for which the luminaire is designed

Description of class	Symbol used to mark luminaires
Luminaires suitable for direct mounting only on non-combustible surfaces	No symbol, but a warning notice is required
Luminaires with or without built-in ballast or transformer suitable for direct mounting on normally flammable surfaces	

Source: Adapted from BS EN 60598

Table 5.3 lists the luminaire classes according to the type of protection against electric shock. Class 0 luminaires are not permitted in the UK by reason of the Electrical Equipment (Safety) Regulations and the Electricity (Factories Act) Special Regulations 1908 and 1944.

The degree of protection the luminaire provides against the ingress of dust and moisture is classified according to the Ingress Protection (IP) System. This system describes a luminaire by a two-digit number in which the first digit classifies the degree of protection against the ingress of solid foreign bodies, from fingers to fine dust, and the second digit classifies the degree of protection against the ingress of moisture. Table 5.4 lists the classes of these two digits.

Self-assessment task 5.2

Give an IP rating for:

(a) An interior louvred luminaire.
(b) An enclosed luminaire suitable for use in a swimming pool.
(c) An outdoor open-type road lighting lantern.
(d) An internally illuminated outdoor bollard.

BS EN 60598 applies to most luminaires intended for use in neutral or hostile environments. It does not apply to many of the luminaires intended for use in hazardous environments, i.e. where there is a risk of fire or explosion. For such applications different standards and certification procedures apply. Detailed guidance on this topic can be found in the CIBSE Lighting Guide: *Hostile and hazardous environments*.

Luminaires in flammable atmospheres

Atmospheres where there is a risk of explosion are classified:

Zone 0   in which an explosive gas–air mixture is continuously present, or present for long periods.

Zone 1   in which an explosive gas–air mixture is likely to occur in normal operation.

Zone 2   in which an explosive gas–air mixture is not likely to occur in normal operation, and if it occurs it will exist only for a short time.

Table 5.3 Classification of luminaires according to the type of protection provided against electric shock

Class	Description	Symbol used to mark luminaires
0[a]	A luminaire in which protection against electric shock relies upon basic insulation; this implies that there are no means for the connection of accessible conductive parts, if any, to the protective conductor in the fixed wiring of the installation – reliance, in the event of a failure of the basic insulation, being placed on the environment	No symbol
I	A luminaire in which protection against electric shock does not rely on basic insulation only, but which includes an additional safety precaution in such a way that means are provided for the connection of accessible conductive parts to the protective (earthing) conductor in the fixed wiring of the installation in such a way that the accessible conductive parts cannot become live in the event of a failure of the basic insulation	No symbol
II	A luminaire in which protection against electric shock does not rely on basic insulation only, but in which additional safety precautions, such as double insulation or reinforced insulation, are provided, there being no provision for protective earthing or reliance upon installation conditions	▢
III	A luminaire in which protection against electric shock relies upon supply at safety extra low voltage (SELV) or in which voltages higher than SELV are not generated. The SELV is defined as a voltage which does not exceed 50 V a.c., r.m.s., between conductors or between any conductor and earth in a circuit which is isolated from the supply mains by such means as a safety isolating transformer or converter with separate windings	◇III

[a] Class 0 luminaires are not permitted in the UK.
Source: Adapted from BS EN 60598.

In Zone 0 there is no acceptable lighting system. Zone 1 permits the use of flameproof equipment certified for use in a specific hazardous situation and 'increased safety protection' (e). Zone 2 permits a lesser degree of protection using non-spark luminaires (N). A brief description of each type is given below.

Pressurised systems (P)

These may be used in both Zones 1 and 2. In one type, air or an inert gas is maintained within the enclosure at a pressure sufficient to prevent ingress of the surrounding, possibly flammable, atmosphere to the enclosure. In another, a flow of air or of an inert gas is maintained to sweep away any flammable vapour that may enter the luminaire.

Table 5.4 The degrees of protection against the ingress of solid bodies (first characteristic numeral) and moisture (second characteristic numeral) in the Ingress Protection (IP) System of luminaire classification

First characteristic numeral	Degree of protection	
	Short description	Brief details of objects which will be 'excluded' from the enclosure
0	Non-protected	No special protection
1	Protected against solid objects greater than 50 mm	A large surface of the body, such as a hand (but no protection against deliberate access). Solid objects exceeding 50 mm in diameter
2	Protected against solid objects greater than 12 mm	Fingers or similar objects not exceeding 80 mm in length. Solid objects exceeding 12 mm in diameter
3	Protected against solid objects greater than 2.5 mm	Tools, wires, etc., of diameter or thickness greater than 2.5 mm. Solid objects exceeding 2.5 mm in diameter
4	Protected against solid objects greater than 1.0 mm	Wires or strips of thickness greater than 1.0 mm. Solid objects exceeding 1.0 mm in diameter
5	Dust protected	Ingress of dust is not totally prevented but dust does not enter in sufficient quantity to interfere with satisfactory operation of the equipment
6	Dust-tight	No ingress of dust
Second characteristic numeral	Degree of protection	
	Short description	Details of the type of protection provided by the enclosure
0	Non-protected	No special protection
1	Protected against dripping water	Dripping water (vertically falling drops) shall have no harmful effect
2	Protected against dripping water when tilted up to 15°	Vertically dripping water shall have no harmful effect when the enclosure is tilted at any angle up to 15° from its normal position
3	Protected against spraying water	Water falling as a spray at an angle up to 60° from the vertical shall have no harmful effect
4	Protected against splashing water	Water splashed against the enclosure from any direction shall have no harmful effect
5	Protected against water jets	Water projected by a nozzle against the enclosure from any direction shall have no harmful effect

Table 5.4 cont'd

Second characteristic numeral	Degree of protection	
	Short description	Details of the type of protection provided by the enclosure
6	Protected against heavy seas	Water from heavy seas or water projected in powerful jets shall not enter the enclosure in harmful quantities
7	Protected against the effects of immersion	Ingress of water in a harmful quantity shall not be possible when the enclosure is immersed in water under defined conditions of pressure and time
8	Protected against submersion	The equipment is suitable for continuous submersion in water under conditions which shall be specified by the manufacturer. *Note:* Normally, this will mean that the equipment is hermetically sealed. However, with certain types of equipment it can mean that water can enter but only in such a manner that it produces no harmful effects

Source: Adapted from BS EN 60598.

Both types require a system of interconnected enclosures with provision for switching off the electrical supply automatically in the event of failure to maintain the air pressure or flow.

A flameproof enclosure

This term is applied to luminaires certified by BASEEFA (British Approvals Service for Electrical Equipment in Flammable Atmospheres) and complying with BS 4683 Part 2 or the earlier standards BS 229 *Flameproof enclosures* and BS 889 *Flameproof electric lighting fittings*, at present under revision. It is defined as

able to withstand an explosion of the flammable gas or vapour which may enter it without suffering damage and without communicating the internal flammation to the external flammable gas or vapour for which it is designed through any joints or structural openings in the enclosure.

It is necessary to control the length of path and gap width to cool the products of combustion and to prevent transmission from the inside to the outside surrounding atmosphere.

Increased safety or type of protection (e)

Until recently, only flameproof or pressurised luminaires were allowed in Zone 1 areas; now 'increased safety' type of protection (e) is permitted. This type of

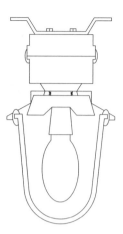

Fig. 5.13   Typical type 'N' luminaire for Zone 2 areas (courtesy Thorn Lighting Ltd)

protection is covered by IEC recommendations. The protection is defined in BS 4683 Part 4 as follows:

> A method of protection by which additional measures are applied to electrical equipment so as to give increased security against the possibility of excessive temperatures and of the occurrence of arcs and sparks during the service life of the apparatus. It applies to electrical equipment, no part of which produce arcs, sparks or exceed the limiting temperatures in normal service.

The light sources allowed are cold starting, fluorescent tubular lamps with single-pin caps, tungsten-filament lamps for general lighting service and mixed-light (i.e. tungsten/mercury) lamps. The luminaire enclosure must also meet minimum strength requirements.

Type (e) luminaires may be used in Zone 1 and Zone 2 areas if the mechanical protection is at least equal to IP 54. The protection type (e) is suitable for use in all gases and vapours, as far as explosion risk is concerned, if the temperature class is acceptable and the materials of construction are compatible with vapours etc. in the surrounding environment.

Non-sparking (N)

This class of luminaire is suitable for Zone 2 areas. Type of protection (N) is defined in BS 4683 Part 3 as

> a type of protection applied to electrical apparatus such that, in normal operation, it is not capable of igniting a surrounding explosive atmosphere and a fault capable of causing ignition is not likely to occur.

It probably covers up to 80 per cent of hazardous areas likely to be encountered. An outline sketch of a typical class (N) luminaire is shown in Fig. 5.13.

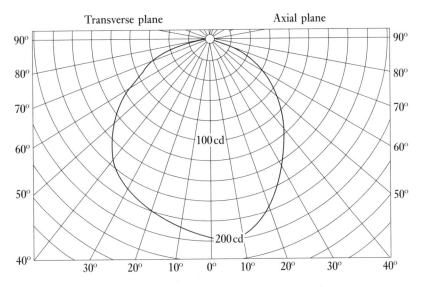

Fig. 5.14 Intensity distribution of a twin-lamp reflector-type luminaire

Photometric performance of luminaires

When selecting suitable luminaires for lighting schemes, many manufacturers' catalogues only provide a picture, some dimensions, lamp sizes and perhaps the price. Other manufacturers may provide complete photometric data, and when carrying out any lighting design calculations it is sensible to base any predictions on luminaires of known performance. The requirements will be considered further in Chapter 7 but at this stage it is useful to consider the minimum information needed for lighting design calculations.

Light intensity distribution

This can be in polar diagram form. Figure 5.14 shows a typical fluorescent luminaire distribution. The right-hand curve is for the vertical plane through the lamp axis and the left-hand curve is for the vertical plane across the lamp axis (transverse plane). The candela scale should be in relation to a lamp output of 1000 lm. This value should then be adjusted for the output of the actual lamp (or lamps).

Example

The data in Fig. 5.14 apply to a twin-lamp reflector-type luminaire. What would be the intensity at 45° in the transverse plane if the lamps were two 58 W white fluorescent tubes?

Solution

The lumen output of the lamp can be found in Appendix C. It is 4600 lm.

The transverse intensity at 45° = 150 cd

$$\text{The correction factor} = \frac{2 \times 4600}{1000}$$

$$\text{Hence, the actual intensity} = 150 \times \frac{2 \times 4600}{1000} \text{ cd}$$

$$= 1380 \text{ cd}$$

Light output ratios

These, as already explained in Chapter 2, are a measure of the light efficiency. All three values are normally quoted, i.e. DLOR, ULOR and LOR. Combining light intensity distribution values and light output ratios it is fairly simple for a manufacturer to produce utilisation factor tables. This, again, is part of Chapter 7, but the data are available in CIBSE Technical Memorandum No. 5 to produce these tables.

Spacing to height ratios

When planning a scheme, the ratio of the mounting height of the luminaires above the working plane to the spacing between luminaires should not exceed a certain value if the lighting is to be reasonably uniform. This ratio is given by the manufacturers and discussed in more detail in Chapter 7.

Luminance distribution

This comprises a table of luminance against angle of view. The information may be needed when calculating discomfort glare and these data are again based on 1000 lm and need correction.

Production of photometric data

Data sheets, as illustrated in Tables 7.3 and 7.10 of Chapter 7, are all derived from the intensity distributions of the luminaire, measured using an intensity distribution photometer. This is basically a photometric device that can rotate a photocell around the luminaire. It measures the illuminance which, using eqn [2.7], is converted to intensity. Most photometers are large because, for eqn [2.7] to be applied, the photometric distance should be at least five times the length of the source, e.g. for a 1.5 m luminaire the measuring distance is 7.5 m or more.

The method of photometry is set out in BS 5225 and the calculation of the data is set out in CIBSE Technical Memorandum No. 5 for utilisation factors and spacing to height ratios, and the CIBSE Technical Memorandum No. 10, *The calculation of glare indices*.

Manufacturers differ in the way they present these types of data and they may provide one set of data for a range of luminaires where a number of correction factors have to be applied. The CIBSE Lighting Code contains a useful summary of typical luminaire characteristics, and this type of guide enables preliminary planning before a specific manufacturer's luminaire is selected, but final calculations should be used on manufacturers' photometric data, preferably accredited by the BSI.

Self-assessment task 5.3

What is the minimum photometric data you could reasonably expect a manufacturer of standard luminaires to provide?

LUMINAIRE APPEARANCE

It is beyond the scope of the book to discuss the aesthetics of luminaire design, but the very nature of lighting is to attract attention to itself. The most efficiently designed lighting scheme can look dull, inappropriate or quite offensive. The overall impression is a combination of the luminaire and the room appearance, and a lighting designer – be he or she an engineer, an architect or an interior designer – should develop a sense of anticipating what the scheme will look like. Some people have a flair for this, but much can be achieved, or avoided, by looking critically at existing lighting and forming positive opinions about it.

EXERCISES

5.1   Design a luminaire for use in a hospital ward. The design has to meet the following requirements:

(a) The luminance from all angles of view of the staff and patient must not exceed 1000 cd/m^2.
(b) The lamps must be tubular fluorescent.
(c) There must be at least 20 per cent upward light.
(d) There must be a minimum of dust ledges.
(e) Night lighting must be included.

The answer should include a clear sketch of the luminaire, the circuit diagram, and a brief description of materials and finishes used.
5.2   Explain why Category 1 louvres will normally have a lower DLOR than Category 3. (These categories are explained in Chapter 8.)
5.3   Refer to the prism in Fig. 5.11. Calculate the angle of deviation between i_1 and i_2 if the incident angle (i_1) is 30° and the refractive index is 1.59 (polystyrene).

CHAPTER 6

DAYLIGHT

TOPICS COVERED

SOURCE OF DAYLIGHT
Luminance distribution of the sky
Variability of daylight
Variation in daylight illuminance

DAYLIGHT FACTOR
Components of daylight factor
Average daylight factor
Orientation factors
Design of windows
Sunlight in buildings
CIBSE Window Design Guide 1987

The vast majority of buildings contain windows, and for most of the working day the illuminance outside a building far exceeds that within. In considering the lighting of the interior a decision *should* be taken on the relative roles of daylight and electric light. In reality this seldom happens. The electric light will function adequately whether or not daylight is present.

However, as electricity costs have risen, this attitude has been challenged. Can daylight be effectively integrated with electric light? Methods have been devised, mainly by the Building Research Establishment, producing in the 1960s Permanent Supplementary Artificial Lighting of Interiors (PSALI) – a serious, if not particularly successful, attempt to produce integrated lighting design.

Work today is more concerned with economics and the savings that might be achievable. To achieve this it has to be possible to predict, with some degree of accuracy, the lighting levels that daylight will provide within a building throughout the year. This information is of interest both to the architect who designs the windows and to the engineer who designs the services.

SOURCE OF DAYLIGHT

The source of daylight is the sun, emitting energy in a continuous spectrum similar to that of a black-body radiator at a temperature of approximately 6000 K.

This energy does not reach the earth's surface without change, as it is absorbed and scattered by the chemical elements, moisture and dust particles in the atmosphere. The scattering of the sun's energy by the atmosphere produces the blue sky which can then be considered as the effective source of daylight other than direct sunlight.

Luminance distribution of the sky

Before calculations of illuminance due to the sky can be made, the luminance of the sky must be determined. Originally it was assumed that the overcast sky was such a good diffusing medium that the luminance of all parts of the sky was the same, i.e. that the sky had a uniform luminance. This type of sky is known as *uniform sky* and under this condition the horizontal illuminance due to an unobstructed hemisphere of overcast sky is given by

$$E = L \times \pi \text{ lx} \qquad [6.1]$$

where L is the sky luminance in candelas per square metre, as shown in Fig. 6.1.

Measurements of sky luminance show that while the sky has a reasonably uniform luminance in azimuth, the luminance in any vertical plane varies. In 1955 the CIE adopted a non-uniform sky as defined by $L_\theta = \frac{1}{3}L_Z(1 + 2 \sin\theta)$ as standard, and this is known as the CIE Overcast Sky, where L_Z is the luminance of zenith.

This form of sky is generally used for daylight calculations in a temperate climate, such as that in the UK.

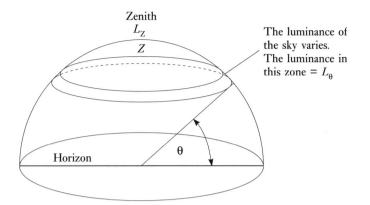

Fig. 6.1 The CIE overcast sky

Variability of daylight

The amount of direct sunlight is so variable and also so unpredictable in countries with climates like that of the UK that most of the calculations of illuminance due to daylight are limited to estimating the amount of light received from an overcast sky. Sunlight is, however, taken into account for estimation of solar heat gain, glare and damage to works of art.

Variation in daylight illuminance

Measurements of horizontal illuminance due to the whole sky, excluding direct sunlight, have been made over a long period at various parts of the UK. Figure 6.2 shows the results of measurements made by the National Physical Laboratory. The maximum value of illuminance is about 35 000 lx and this occurs only in July and for only a relatively short period about noon. A standard outdoor illuminance of 5000 lx forms the basis for daylight calculations for the UK. Thus it is assumed that the average illuminance due to a complete hemisphere of sky – as measured, say, on the roof of a high building – is 5000 lx.

Self-assessment task 6.1

To maintain at least 500 lx from daylight with a daylight factor of 2 per cent, calculate the number of hours this will be achieved throughout the year between 8 a.m. and 6 p.m.

DAYLIGHT FACTOR

An indication of the amount of daylight at a point within a room is the ratio of the daylight illuminance at the point to the instantaneous illuminance outside the building from a complete hemisphere of sky (excluding direct sunlight).

Time of day (LAT)

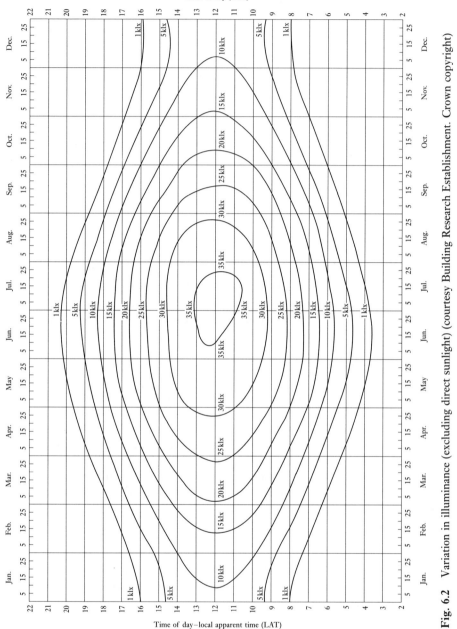

Time of day – local apparent time (LAT)

Fig. 6.2 Variation in illuminance (excluding direct sunlight) (courtesy Building Research Establishment. Crown copyright)

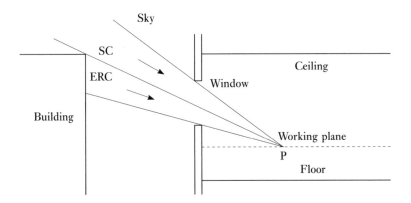

Fig. 6.3   The three components of daylight

This ratio is known as the *daylight factor* (DF), and is usually quoted as a percentage:

$$\text{Daylight factor} = \frac{\begin{array}{c}\text{horizontal illuminance at}\\\text{a point in an interior}\end{array}}{\begin{array}{c}\text{horizontal illuminance at the same}\\\text{instant due to an unobstructed sky}\end{array}} \times 100\% \qquad [6.2]$$

Components of daylight factor

It is desirable at the design stage of the building to predict the amount of daylight that will be obtained for a given window configuration. It is necessary, therefore, to consider how daylight reaches a point within a room and this can be done by dividing the illuminance received into three components, as shown in Figs 6.3 and 6.4. Figure 6.3 shows light reaching the point P directly from the sky. This is known as the sky component (SC) of daylight.

In many cases nearby buildings may obstruct the light from the sky to the point P, and hence reduce the sky component. The surfaces outside the room do, however, reflect light from other parts of the sky into the room and contribute a little towards the daylight within the room, as shown in the figure. This component is known as the *externally reflected component* (ERC) of daylight.

The final component is due to the light entering the room being reflected onto the reference plane, as shown in Fig. 6.4. In this case the window could be considered as an area source emitting light onto all of the room surfaces, some of which is reflected onto the reference plane, increasing the illuminance. This component is known as the *internally reflected component* (IRC) of daylight.

When these components are evaluated as a percentage of the external illuminance due to an unobstructed hemisphere of sky, their arithmetic sum gives the daylight factor. Thus,

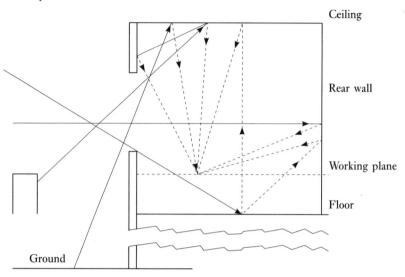

Fig. 6.4   The IRC component of daylight

Daylight factor = sky component + externally reflected component
+ internally reflected component

or

$$DF = SC + ERC + IRC \qquad [6.3]$$

Calculation of the components of daylight factor (DF)

The three components can be calculated by a range of methods. The SC and ERC are both found by considering the geometry of the visible sky or external reflecting surfaces at a point on a horizontal plane within the room. IRC is based on interreflection theory and can be found from a formula, nomogram or tables. The most commonly used methods in the UK are:

- *SC and ERC*: Waldram diagram; BRE protractors; BRE tables; Pilkington 'pepper pot' diagrams.
- *IRC*: Formula; BRE tables; BRE nomograms.

The presentation and explanation of all these methods would require a separate book. This chapter refers only to tables and formulae. This is not because they are the easiest methods, but because they do not require special protractors or overlays. Computer programs are available for these calculations and would be the best choice if a number of calculations were necessary.

Table 6.1 Sky components (CIE standard overcast sky) for vertical glazed rectangular windows

Ratio H/D = Height of window head above working plane/Distance from window

H/D	0.2	0.4	0.6	0.8	1.0	1.2	1.4	1.6	2.0	∞	Angle of obstruction
∞	2.5	4.9	6.9	8.4	9.6	10.7	11.6	12.2	13.0	15.0	90°
	2.4	4.8	6.8	8.3	9.4	10.5	11.1	11.7	12.7	14.2	79°
	2.4	4.7	6.7	8.2	9.2	10.3	10.9	11.4	12.4	13.7	76°
	2.4	4.6	6.6	8.0	9.0	10.1	10.6	11.1	12.2	13.3	74°
	2.3	4.5	6.4	7.8	8.7	9.8	10.2	10.7	11.7	12.7	72°
	2.3	4.5	6.3	7.6	8.6	9.6	10.0	10.5	11.4	12.3	70°
	2.2	4.4	6.2	7.5	8.4	9.3	9.8	10.2	11.1	11.9	69°
	2.2	4.3	6.0	7.3	8.1	9.1	9.5	10.0	10.7	11.5	67°
	2.1	4.1	5.8	7.0	7.9	8.7	9.1	9.6	10.2	10.9	66°
2.0	2.0	4.0	5.6	6.7	7.5	8.3	8.7	9.1	9.7	10.3	63°
	2.0	3.9	5.4	6.5	7.3	8.1	8.5	8.8	9.4	9.9	62°
	1.9	3.8	5.3	6.3	7.1	7.8	8.2	8.5	9.0	9.5	61°
	1.9	3.6	5.1	6.1	6.8	7.5	7.8	8.2	8.6	9.1	60°
	1.8	3.5	4.9	5.8	6.5	7.2	7.5	7.8	8.2	8.6	58°
1.5	1.7	3.3	4.6	5.6	6.2	6.8	7.1	7.4	7.8	8.1	56°
1.4	1.6	3.2	4.4	5.2	5.9	6.4	6.7	7.0	7.3	7.6	54°
1.3	1.5	2.9	4.1	4.9	5.5	5.9	6.2	6.4	6.7	7.0	52°
1.2	1.4	2.7	3.8	4.5	5.0	5.4	5.7	5.9	6.1	6.3	50°
1.1	1.3	2.5	3.4	4.1	4.6	4.9	5.1	5.3	5.4	5.7	48°
1.0	1.1	2.2	3.0	3.6	4.0	4.3	4.5	4.6	4.7	5.0	45°
0.9	0.99	1.9	2.6	3.1	3.4	3.7	3.8	3.9	4.0	4.2	42°
0.8	0.83	1.6	2.2	2.6	2.9	3.1	3.2	3.3	3.3	3.1	30°
0.7	0.68	1.3	1.7	2.1	2.3	2.5	2.5	2.6	2.6	2.8	35°
0.6	0.53	0.98	1.3	1.6	1.8	1.9	1.9	2.0	2.0	2.1	31°
0.5	0.39	0.70	0.97	1.10	1.3	1.4	1.4	1.4	1.5	1.5	27°
0.4	0.25	0.45	0.62	0.75	0.89	0.92	0.95	0.95	0.96	0.98	22°
0.3	0.14	0.26	0.34	0.42	0.47	0.49	0.50	0.50	0.51	0.53	17°
0.2	0.06	0.11	0.14	0.20	0.21	0.22	0.22	0.22	0.23	0.24	11°
0.1	0.02	0.03	0.04	0.05	0.05	0.06	0.06	0.06	0.07	0.08	8°
0	0.2	0.4	0.6	0.8	1.0	1.2	1.4	1.6	2.0	∞	0°

Ratio W/D = Width of window to one side of normal/Distance from window

Source: Building Research Establishment. Crown copyright.

Calculation of sky component (SC)

Table 6.1 is used for determining these factors for windows with clear, vertical, rectangular glazing in conjunction with a CIE standard overcast sky. Other tables are available for other forms of glazing, e.g. roof lights.

The following information is needed:

■ H_1 and H_2, the heights of the window head and cill above the working plane.
■ W_1 and W_2, the distances of the window's vertical edges from a line drawn from the reference point for which the daylight factor is to be calculated, normal to the plane of the window.

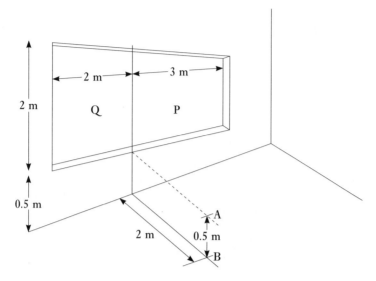

Fig. 6.5

■ *D*, the distance from the reference point to the plane of the window. (This is the plane of the inside of the wall, or the outside, whichever edge of the window aperture limits the view of the sky.)

Example

The SC is required at point A in Fig. 6.5.

Solution

Consider two separate window areas P and Q. In this example the plane of measurement is level with the window cill. *D* is 2 m, H_1 is 2.5 m, H_2 is 0.5 m, W_1 is 3 m and W_2 is 2 m. In this case $H = H_1 - H_2$.

For window area P:

$$\frac{H}{D} = \frac{2.5 - 0.5}{2}$$

$$= 1$$

$$\frac{W}{D} = \frac{3}{2}$$

$$= 1.5$$

From Table 6.1 the SC is 4.55 per cent.

For window area Q:

$$\frac{H}{D} = \frac{2.5 - 0.5}{2}$$

$$= 1$$

$$\frac{W}{D} = \frac{2}{2}$$
$$= 1$$

Hence, from Table 6.1 the SC is 4 per cent.

$$\therefore \quad \text{Total SC} = 8.55 \text{ per cent}$$

The point A, however, may not be at the same height as the cill.

Example

The SC is required at floor level point B.

Solution

It is necessary to think of the window extending down to floor level, calculate the SC, and then deduct the SC for the bit of 'extra' window between the cill and the floor.

For area P down to ground level:

$$\frac{H}{D} = \frac{2.5}{2}$$
$$= 1.25$$

$$\frac{W}{D} = 1.5 \quad \text{(as in previous example)}$$

For area Q down to ground level:

$$\frac{H}{D} = 1.25$$

$$\frac{W}{D} = 1 \quad \text{(as in previous example)}$$

Hence for this enlarged area:

$$SC = 6.05 + 5.25$$
$$= 11.3 \text{ per cent}$$

Now subtract the 'extra' window below P:

$$\frac{H}{D} = \frac{0.5}{2}$$
$$= 0.25$$

$$\frac{W}{D} = 1.5$$

and below Q:

$$\frac{H}{D} = 0.25$$

$$\frac{W}{D} = 1$$

$$\therefore \quad SC = 11.3 - 0.7$$
$$= 10.6 \text{ per cent}$$

It is interesting to note in these examples that although A is nearer the window than B, the SC is less. This is because the measurement is on a horizontal plane and at A very little light is seen or received from the lower part of the window. At B the window is more effective. Window height can be more significant than window width.

Calculation of externally reflected component (ERC)

If the direct entry of light through the window is severely limited by an external obstruction, it will be necessary to calculate the ERC. This can also be done using Table 6.1. The procedure is to treat the external obstruction visible from the reference point as a patch of sky whose luminance is some fraction of the sky obscured. In other words, the SC for the obstructed area is first calculated as described above and is then converted to the ERC by multiplying by the ratio of the luminance of the obstructed area to the sky luminance.

This is one of those calculations which is 'easier said than done'. The difficulty is to establish how much of the sky is obstructed, as seen from the point of measurement. Also, it is necessary to judge the reflectance of the obstruction. If it is 0.2 then the calculated ERC will be 0.2 of the equivalent SC.

Example

If in Fig. 6.5 the view from A shows a sky obstruction for the lower half of the window and the obstruction has a reflectance of 0.1, what percentage of sky and reflected sky reaches A?

Solution

The window must be considered as two sources. Taking the lower half, it effectively transmits light from a sky 0.1 the luminance of the real sky.

The SC for the lower half is calculated in the following way.

For window area P:

$$H = 1.5 - 0.5$$
$$\therefore \quad \frac{H}{D} = \frac{1.0}{2}$$
$$= 0.5$$
$$\frac{W}{D} = 1.5 \quad \text{(as before)}$$

For window area Q:

$$\frac{H}{D} = 0.5$$
$$\frac{W}{D} = 1 \quad \text{(as before)}$$

Hence

SC = 1.4 + 1.6
    = 3.0 per cent

This is for an open-sky view, but as the view is obstructed,

ERC = SC × reflectance
    = 3.0 × 0.1
    = 0.3 per cent

This is a small amount as might be expected in this situation, and illustrates how significantly external obstructions can reduce available daylight.

The SC will be for the top half of the window, but nearly all necessary calculations have already been done.

SC = SC for whole unobstructed window *minus* SC for lower half *if* unobstructed
   = 8.55 − 3.0
   = 5.55 per cent

If the window was totally obstructed

ERC = 8.55 × 0.1
    = 0.8 per cent

Therefore, this makes the daylight of little value in terms of illumination.

'No-sky' line

If there are any significant external obstructions, a line can be drawn on the working plane in the room indicating the points at which direct sky view ceases. Such a line is shown in Fig. 6.6. When considering the lighting value of daylight it is suggested that the area beyond the 'no-sky' line will receive inadequate daylight for normal working illumination.

Calculation of internally reflected component (IRC)

If all the room surfaces were black, the daylight received at a point in the room would be that which comes direct from the window. In practice, all surfaces reflect light and this reflected component can be significant. The calculation of the reflected component would be a tedious process but for the tables available.

Table 6.2 was formulated by the BRE to give minimum internally reflected components of daylight factor, assuming a CIE standard overcast sky and a known scheme of decoration. The table was designed primarily for rooms 6 m square (i.e. 36 m² in floor area) and 3 m ceiling height, having a window (glazed with ordinary glass) on one side extending from a 1 m cill to the ceiling, but by means of simple conversion factors (Table 6.3) rooms of 10–100 m² floor area, with ceiling heights ranging from 2.5 to 4 m, can be examined. Initially, the ceiling reflection factor is assumed to be 70 per cent, but other values can be allowed for by means of conversion factors. A long external obstruction of 20° measured from the centre of

Table 6.2 The minimum IRC of DF (per cent)[a]

Ratio of window area to floor area	Window area as percentage of floor area	Floor reflection factor											
		10%				20%				40%			
		Wall reflection factor				Wall reflection factor				Wall reflection factor			
		20%	40%	60%	80%	20%	40%	60%	80%	20%	40%	60%	80%
1:50	2	—	—	0.1	0.2	—	0.1	0.1	0.2	—	0.1	0.2	0.2
1:20	5	0.1	0.1	0.2	0.4	0.1	0.2	0.3	0.5	0.1	0.2	0.4	0.6
1:14	7	0.1	0.2	0.3	0.5	0.1	0.2	0.4	0.6	0.2	0.3	0.6	0.8
1:10	10	0.1	0.2	0.4	0.7	0.2	0.3	0.6	0.9	0.3	0.5	0.8	1.2
1:6.7	15	0.2	0.4	0.6	1.0	0.2	0.5	0.8	1.3	0.4	0.7	1.1	1.7
1:5	20	0.2	0.5	0.8	1.4	0.3	0.6	1.1	1.7	0.5	0.9	1.5	2.3
1:4	25	0.3	0.6	1.0	1.7	0.4	0.8	1.3	2.0	0.6	1.1	1.8	2.8
1:3.3	30	0.3	0.7	1.2	2.0	0.5	0.9	1.5	2.4	0.8	1.3	2.1	3.3
1:2.9	35	0.4	0.8	1.4	2.3	0.5	1.0	1.8	2.8	0.9	1.5	2.4	3.8
1:2.5	40	0.5	0.9	1.6	2.6	0.6	1.2	2.0	3.1	1.0	1.7	2.7	4.2
1:2.2	45	0.5	1.0	1.8	2.9	0.7	1.3	2.2	3.4	1.2	1.9	3.0	4.6
1:2	50	0.6	1.1	1.9	3.1	0.8	1.4	2.3	3.7	1.3	2.1	3.2	4.9

[a] Assuming ceiling reflection factor = 70 per cent; angle of external obstruction = 20 degrees.

Source: Building Research Establishment. Crown copyright.

Table 6.3 Conversion factors to Table 6.2 for different floor areas and ceiling reflectances

Floor area (m²)	Wall reflection factor			
	20%	40%	60%	80%
10	0.6	0.7	0.8	0.9
100	1.4	1.2	1.0	0.9

Source: Building Research Establishment. Crown copyright.

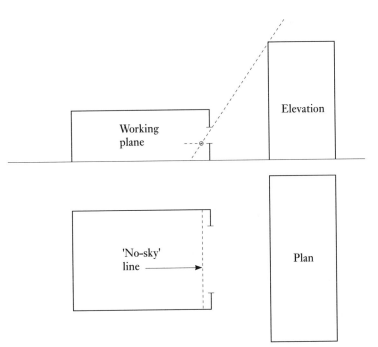

Fig. 6.6 The 'no-sky' line due to external obstructions

the window is assumed as a common condition. In cases where there is more than one window in the room, the internally reflected components should be calculated separately for each window and then they are added together, remembering that the minimum IRC for each window occurs at points in the room farthest from it.

Example

The room under examination is 6 m × 6 m (i.e. 36 m²) × 3 m high and has a window in one wall with an area of 8.2 m². The floor has an average reflectance of 20 per cent; therefore, ratio of window area to floor area = 1 : 4.4.

Table 6.4   Conversion factors for average internally reflected component

Wall reflectance (%)	Conversion factor
20	1.8
40	1.4
60	1.3
80	1.2

Source: Building Research Establishment. Crown copyright.

Table 6.5   Light losses due to glazing bars

For large-paned windows	Reduction factor
All metal windows	0.8
Metal windows in wood frames	0.75
All wood windows	0.65–0.70

Source: Building Research Establishment. Crown copyright.

The walls = 40 per cent reflectance
The ceiling = 70 per cent reflectance

Referring to Table 6.2 it will be seen that the minimum IRC of DF lies between 0.6 and 0.8, or 0.7 per cent.

The conversion factor for ceiling reflectance is approximately 0.1 for 10 per cent change in reflectance below 70 per cent, e.g. for 50 per cent the reflection factor is $1.0 - 0.2 = 0.8$.

The above method calculates the *minimum* value. It is usual practice to include this value in DF calculations even though the value of IRC will increase as the point of measurement moves closer to the window. When calculating the average DF, the average IRC can be found using the conversions in Table 6.4.

Light losses

Allowance has to be made for light lost due to obstruction by window bars, transmission through the glass and absorption due to dirt on the glass and room surfaces.

1. *Glazing bars.* Full window opening is assumed, and the reduction factors given in Table 6.5 should be applied.
2. *Glass transmission loss.* Allowance is made in the calculations for normal single sheet window losses. There are certain tinted glasses used for glare reduction and heat–absorbing glasses for reduction in solar gain. A range of transmission factors is given in Table 6.6. Where double glazing is used, a factor of 0.9 should be applied.

Table 6.6 Relative transmittance and correction factors for various glazing materials

Material	Diffuse transmittance	Recommended correction factor to apply to daylight factor *Note*: Correction is proportional to transmission through clear glass
Transparent glasses		
6 mm clear float	0.85	1.0
6 mm polished wired	0.80	0.95
Patterned and diffusing glasses		
3 mm rolled	0.80 to 0.85	0.95
6 mm rough cast	0.80 to 0.85	0.95
6 mm wired cast	0.75	0.90
Special glasses		
Heat-absorbing tinted float	0.45 to 0.75	0.50 to 0.90
Heat-absorbing tinted cast	0.50 to 0.65	0.60 to 0.75
Laminated insulating[a]	0.20 to 0.60	0.25 to 0.70
Plastics sheets		
Corrugated resin-bonded		
Glass-fibre-reinforced		
Roofing sheets		
moderately diffusing	0.75[a] to 0.80	0.90
heavily diffusing	0.66[a] to 0.75	0.75 to 0.90
very heavily diffusing	0.55[a] to 0.70	0.65 to 0.80
Diffusing plain opal 3 mm acrylic sheets	0.55 to 0.78 (depending on grade)	0.65 to 0.90

[a] Typical measurement values.
Source: Building Research Establishment. Crown copyright.

3. *Maintenance factors.* Tables 6.7 and 6.8 give the factors as recommended in BRE Digest 42 to allow for dirt on the windows and on the room surfaces.

Applying these factors to the calculations,

$$\text{Corrected DF} = (\text{SC} + \text{ERC}) \times \text{correction for glazing bars}$$
$$\times \text{ correction for type of glazing} \times \text{correction for}$$
$$\text{dirt on glazing} + (\text{IRC}) \times \text{above corrections}$$
$$\times \text{ correction for dirt on room surfaces} \qquad [6.4]$$

Example

Returning to the first example and assuming the IRC is 0.5 per cent, the window frames are metal, the windows are clear and double glazed, and the room is an office in an industrial area, then

Table 6.7   Maintenance factors to be applied to the total DF or to each of its three components to allow for dirt on the glazing

Location of building	Inclination of glazing	Maintenance factor	
		Non-industrial or clean industrial work	Dirty industrial work
Non-industrial or clean industrial area	Vertical	0.9	0.8
	Sloping	0.8	0.7
	Horizontal	0.7	0.6
Dirty industrial area	Vertical	0.8	0.7
	Sloping	0.7	0.6
	Horizontal	0.6	0.5

Source: Building Research Establishment. Crown copyright.

Table 6.8   Maintenance factors to be applied to the IRC of DF to allow for dirt on room surfaces

Location of building	Maintenance factor	
	Non-industrial or clean industrial work	Dirty industrial work
Non-industrial or clean industrial area	0.9	0.7
Dirty industrial area	0.9	0.6

Source: Building Research Establishment. Crown copyright.

$$\text{Uncorrected DF} = \text{SC} + \text{ERC} + \text{IRC}$$
$$= 08.55 + 0 + 0.5$$
$$= 9.05 \text{ per cent}$$

Corrections:

$$\text{Metal glazing} = 0.8$$
$$\text{Double glazing} = 0.9$$
$$\text{Dirt on windows} = 0.8$$
$$\text{Dirt in room} = 0.8$$
$$\therefore \quad \text{Corrected DF} = (8.55 \times 0.8 \times 0.9 \times 0.8) + (0.5 \times 0.8 \times 0.9 \times 0.8 \times 0.8)$$
$$= 5.1 \text{ per cent}$$

Note that the corrections for a fairly normal situation have almost halved the DF.

Average daylight factor

For rooms with side windows the daylight falls rapidly as the distance from the window is increased. Figure 6.7 shows the fall-off for a typical side-lit room. Despite this variation, an average value can be calculated. This may appear to be

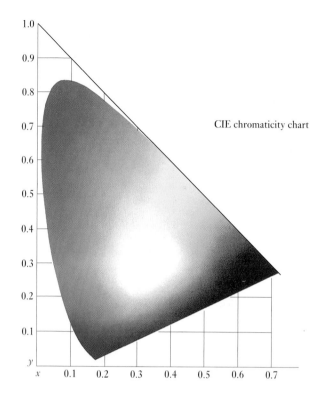

CIE chromaticity chart

Plate 1 The CIE chromaticity diagram

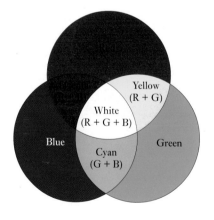

Yellow (R + G)

White (R + G + B)

Blue

Cyan (G + B)

Green

Plate 2 Mixture of three coloured patches of light – additive mixing

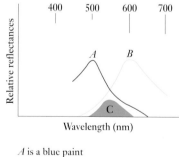

A is a blue paint
B is a yellow paint
C is the mix of A and B and is green

Plate 3 Mixture of yellow and blue pigment – subtractive mixing

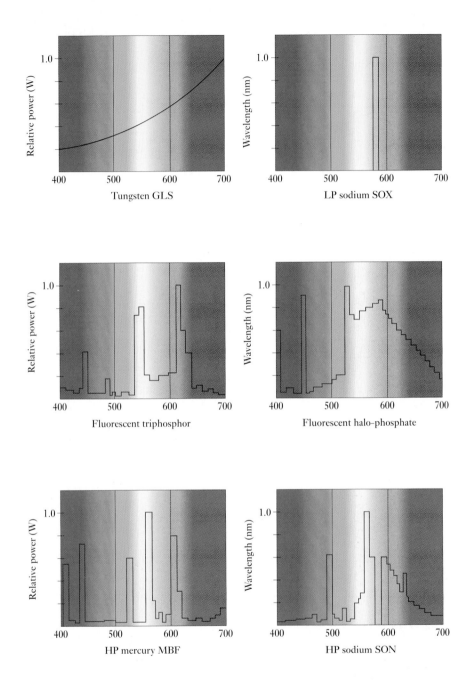

Plate 4 Typical light spectra

Plate 5 Don Valley Stadium, Sheffield © Thorn Lighting Limited

Plate 6 Eilean Donan Castle © The Highland Council

Plate 7 The Bentall Centre, Kingston-upon-Thames © Dennis Gilbert/VIEW

Plate 8 Windsor Leisure Pool © Windsor and Maidenhead Council

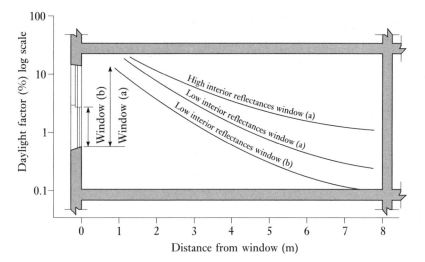

Fig. 6.7 Typical plots of DF against room depth for various window heights and reflectances

meaningless, but work by Lynes, Longmore and others suggests that it can give an overall guide to the adequacy of the daylight.

The CIBSE Lighting Code suggests that when the average DF exceeds 5 per cent in a building which is used mainly during the day, electricity consumption for lighting should be too small to justify elaborate control systems on economic grounds, provided that switches are sensibly located. When the average DF is between 2 and 5 per cent, the electric lighting should be planned to take full advantage of available daylight. Localised or local lighting may be particularly advantageous, using daylight to provide the general lighting. When the average DF is below 2 per cent, supplementary electric lighting will be needed almost permanently.

The average DF will often give the designer sufficient information on which to base decisions on the relationship between natural and electric lighting. However, where more detailed information on the DF is necessary, the methods for point-by-point DF can be used.

A number of formulae are suggested for the calculation of the average DF. The simplest is known as the Littlefair/Plymouth expression, but is only suitable for broad assessments:

$$\text{Average DF} = \frac{\tau W \theta}{A(1 - R^2)} \qquad [6.5]$$

where τ is the diffuse transmittance of glazing material, W the net glazed area of window, A the area of all surfaces, R the average reflectance of all room surfaces and θ the angle subtended, in the vertical plane normal to the window, by sky visible from the centre of the window. (This is shown in Fig. 6.8, and is quoted in degrees.)

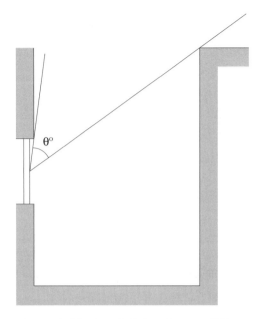

Fig. 6.8 The obstruction angle θ in the calculation of average DF

Self-assessment task 6.2

Taking the room dimensions from the example on p. 135 calculate the average daylight factor. Assume clear double glazing and no external obstructions, and a floor reflectance of 20 per cent.

Orientation factors

The number of variables in daylight calculations can give rise to large errors. However, this is no excuse for oversimplification and approximation. One aspect under research is the error implicit in assuming a standard sky condition independent of building orientation. It is reasonable to assume that, even with overcast conditions, windows facing the direction of the sun will receive more light than those facing away. The CIBSE Window Design Guide contains a set of orientation factors (Table 6.9) which should be used when calculating the DF for energy-saving estimates, i.e.

Orientation weighted DF = orientation factor × overcast sky DF   [6.6]

Design of windows

By considering windows solely as admitters of daylight it may seem that the achievement of a certain minimum DF is the main design criterion. Electric

Table 6.9 Orientation factors for use with overcast sky daylight factors

Direction	Orientation factor
South-facing window	1.20
East-facing window	1.04
West-facing window	1.00
North-facing window	0.77

Source: Chartered Institution of Building Services. CIBSE Window Design Guide.

lighting is today often used during daylight hours and it may be the case that in trying to achieve a minimum DF of, say, 2 per cent, the windows become too large causing more serious glare and heating problems. It may be preferable to design to a lower DF and integrate the daylight with the electric light.

Figure 6.9 shows how the DF varies across a room using two types of window, both having the same total area: (a) is for three tall windows, and (b) is for one long, high-level window. Window (a) will give a good open view, but (b) will provide more even daylight illumination and a higher minimum DF.

For roof lights, as shown in Fig. 6.10, the daylighting is evenly spread over the working area and much higher factors are normally obtained. These factors are often drastically reduced by overhead obstructions and poor window maintenance.

To achieve an illuminance of 500 lx requires an average DF of around 10 per cent, which is much higher than could reasonably be expected in practice.

While DF is a useful guide to the light penetration from windows into rooms, it should not be the governing factor in window design if there is no economic justification for the large areas of glazing that might be involved.

J. A. Lynes has formulated a design process to establish a window design to achieve a required average DF. This is published in *Lighting Research and Technology* 11.102.79. This calculation is only relevant if the daylight contribution is significant, and to achieve this he suggests:

- No significant part of the working plane shall lie beyond the 'no-sky' line.
- $(L/W + L/H)$ shall not exceed $2/(1 - R_B)$, where L is the depth of room from window to back wall, W the width of room, measured parallel to window, H the height of window head above floor level, and R_B the average reflectance of the half of the interior remote from the window.

If these conditions cannot be satisfied the daylighting will be unsatisfactory whatever the size of the window as the back of the room will always look gloomy compared with the space by the window.

Calculation of daylight factor for roof lights

It is normally assumed that if the light-transmitting structure of the roof is evenly spaced, then the daylighting at floor or working plane level will be reasonably

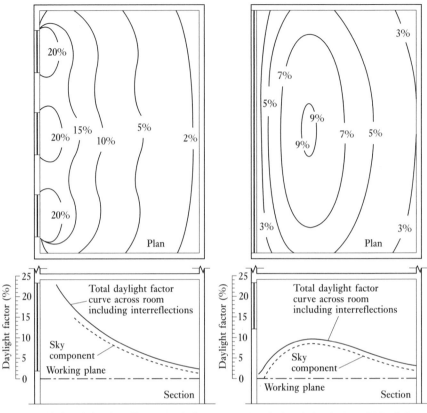

(a) Window design for tall, narrow windows   (b) Window design for long, high windows

Fig. 6.9   The effect of window shape on DF

uniform (see Fig. 6.10). A single calculation is then sufficient to determine the average DF.

A formula to find this value is

$$\text{DF} = 100 \times \text{UF} \times M \times B \times G \times k \text{ per cent} \qquad [6.7]$$

where UF is the utilisation factor of the roof lights as given in Table 6.10 for various reflectances and room indices, and

$$\text{Room index} = \frac{\text{length} \times \text{width of room}}{(L + W) \times \text{roof height above working plane}}$$

M is a light loss factor for dirt on the glass (Table 6.7), B is a correction factor for glazing bars and internal obstructions (Table 6.5), G is a correction factor for the type of glazing (Table 6.6) and k is the ratio of glazing area to floor area.

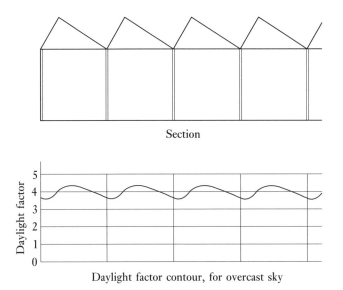

Section

Daylight factor contour, for overcast sky

Fig. 6.10 The distribution of daylight across the floor for a roof-lit interior

Table 6.10 Example of utilisation factors for roof lights (twin monitor)

	Reflectance								
Ceiling	70%			50%			30%		0
Walls	50%	30%	10%	50%	30%	10%	30%	10%	0
Room index	Utilisation factor								
0.6	0.07	0.05	0.04	0.06	0.05	0.04	0.05	0.04	0.03
0.8	0.09	0.07	0.06	0.09	0.07	0.06	0.07	0.06	0.05
1.0	0.12	0.10	0.08	0.11	0.09	0.08	0.09	0.08	0.07
1.25	0.14	0.12	0.10	0.13	0.11	0.10	0.11	0.10	0.09
1.5	0.15	0.13	0.12	0.15	0.13	0.12	0.13	0.11	0.11
2.0	0.17	0.15	0.14	0.16	0.15	0.14	0.15	0.13	0.13
2.5	0.18	0.17	0.15	0.18	0.16	0.15	0.16	0.15	0.14
3.0	0.20	0.18	0.17	0.19	0.18	0.17	0.17	0.16	0.16
4.0	0.21	0.20	0.19	0.20	0.19	0.19	0.19	0.18	0.17
5.0	0.22	0.21	0.20	0.21	0.20	0.19	0.20	0.19	0.18
∞	0.25	0.25	0.25	0.25	0.25	0.25	0.24	0.24	0.23

Source: Pilkington Glass Ltd.

Example

A factory bay is 20 m × 10 m with a twin monitor roof of average height 6 m above the floor. If the reflectances are 0.7 for the ceiling and 0.3 for the walls and k is 0.2, calculate the average DF at floor level. (The glass is 6 mm wired cast.)

Solution

Some assumptions will have to be made concerning the location and obstructions.
To find UF:

$$\text{Room index} = \frac{10 \times 20}{(10 + 20) \times 6}$$
$$= 1.1$$

So from Table 6.10, the reflectances 70 per cent and 30 per cent and a twin monitor roof:

$$\text{UF} = 0.11$$

From Table 6.7 for sloping glazing and clean industrial work and area:

$$M = 0.8$$

From Table 6.5 for metal glazing bars and assuming no internal obstructions:

$$B = 0.8$$

From Table 6.6 for 6 mm wired cast glass:

$$G = 0.9$$

k is given as 0.2. Therefore,

$$\text{DF} = 100 \times 0.11 \times 0.8 \times 0.8 \times 0.9 \times 0.2$$
$$= 1.3 \text{ per cent}$$

Sunlight in buildings

All calculations have referred to an overcast sky and have excluded direct sunlight.
It is not feasible to consider direct sunlight as a positive factor when designing
interior lighting. Much as it is welcomed in home interiors, considerable effort is
taken to exclude it from commercial and industrial buildings.

The exclusion can take various forms including blinds, solar reflecting or
absorbing glasses, or louvres. All these will modify the DF and often nullify the
lighting value of the glazing.

Any detailed discussion on sunlight calculations is outside the scope of this book.
The geometry of sunlight penetration is explained in such documents and books as
BS DD67: 1980 *Basic data for the design of buildings: Sunlight* and *Windows and the
Environment* (Pilkington Advisory Service).

CIBSE Window Design Guide 1987

This guide attempts to coordinate existing thinking on windows and their role in
the building environment, visually, thermally and acoustically. The main text deals
with the process of window design as a transmitter of daylight into buildings. The
appendix, which is longer than the design section, brings to one publication current
data on prediction of daylight and sunlight and other environment topics.

EXERCISES

6.1 Explain the difference between sky factor and daylight factor.

An office is 10 m wide by 6 m deep. The ceiling height is 3 m, the windows are clear double glazed with metal frames facing east and occupy the whole of one 10 m wall. The room reflectances are 70 per cent ceiling, 40 per cent walls and 20 per cent floor. The outside view is unobstructed. Calculate the minimum daylight factor at floor level.

6.2 Calculate the average daylight factor for this office and comment on your answer.

6.3 Discuss critically the statement that 'Daylight is the most expensive way of lighting an interior'.

6.4 Sketch typical daylight factor contours in an interior for the following:

(a) tall narrow windows

(b) windows with external view obstructed

(c) rooms with dark interior decorations

CHAPTER 7

DESIGN OF GENERAL LIGHTING SCHEMES

TOPICS COVERED

HOW MUCH LIGHT?
Designing a general layout
Planning the layout
Types of layout
Relative luminance of room surfaces
Calculation of glare
Disability glare
Discomfort glare
CIE Unified Glare Rating (UGR)
CIE luminance curve system

UPLIGHTING
Lumen method
Thorn Lighting method

EMERGENCY LIGHTING
Planning the layouts
Maintained operation
Non-maintained operation
Location of exit signs (Pr EN1828)

This should be the culmination of all the knowledge gained from the previous chapters. Unfortunately, this is where many people responsible for the design of lighting start and finish. Good design is the application of the best or most appropriate equipment in an economical but effective manner.

This chapter covers the normal design processes for establishing a satisfactory illuminance on the working plane, provided in a glare-controlled manner. The overall pattern of room illuminances is considered in relation to suggestions contained in the CIBSE 1994 Lighting Code. Finally, the requirements for emergency lighting are outlined. The chapter contains a limited amount of photometric data, sufficient to indicate how the data are used, but many manufacturers will usually provide full data for their range of luminaires.

HOW MUCH LIGHT?

Maintained values of illuminance (lux) are specified in the CIBSE 1994 Lighting Code for all the usual locations and occupations. These values are necessary to achieve:

■ Satisfactory illuminance of the task.
■ An agreeable general appearance of the interior.

They are quoted as 'maintained illuminances', normally on a horizontal working plane. The 'standard maintained illuminance' is the value recommended for assumed standard conditions of the lighting application. These conditions include uniformity and anticipated depreciation due to the lamps ageing and the installation becoming dirty. Table 7.1 shows the illuminances relevant to a range of different visual tasks.

This table is expanded in the CIBSE document to quote recommended maintained illuminance values covering over 300 activities or types of interiors.

The illuminance values serve as a guide to good practice. They are not mandatory, although some authorities may adopt them as part of their standard requirements. It is also recognised that there can be circumstances for a given situation where the value should be modified to give a design maintained value, i.e. one applicable to the specific circumstances.

This is demonstrated in the form of a flow chart, and is reproduced in Table 7.2.

The term 'maintained' implies that, provided the installation is cleaned at the specified interval and the lamps changed by the specified hours of use, the average illuminance level for which the lighting has been designed will be maintained. To summarise the above:

■ Standard maintained illuminance (SMI) – a standard scale of values based on specific visual conditions (Table 7.1).
■ Recommended maintained illuminance (RMI) – the application of Table 7.1 to specific situations, e.g. the RMI in the CIBSE general lighting schedule for libraries is 300 lx.

Table 7.1 Examples of activities/interiors appropriate for each maintained illuminance

Standard maintained illuminance (lx)	Characteristics of the activity/interior	Representative activities/ interiors
50	Interiors used rarely with visual tasks confined to movement and casual seeing without perception of detail	Cable tunnels, indoor storage tanks, walkways
100	Interiors used occasionally with visual tasks confined to movement and casual seeing calling for only limited perception of detail	Corridors, changing rooms, bulk stores, auditoria
150	Interiors used occasionally with visual tasks requiring some perception of detail or involving some risk to people, plant or product	Loading bays, medical stores, switchrooms, plant rooms
200	Continuously occupied interiors, visual tasks not requiring perception or detail	Foyers and entrances, monitoring automatic processes, casting concrete, turbine halls, dining rooms
300	Continuously occupied interiors, visual tasks moderately easy, i.e. large details (> 10 minutes arc) and/or of high contrast	Libraries, sports and assembly halls, teaching spaces, lecture theatres, packing
500	Visual tasks moderately difficult, i.e. details to be seen are of moderate size (5–10 minutes of arc) and may be of low contrast. Also colour judgement may be required	General offices, engine assembly, painting and spraying, kitchens, laboratories, retail shops
750	Visual tasks difficult, i.e. details to be seen are small (3–5 minutes of arc) and of low contrast. Also good colour judgements may be required	Drawing offices, ceramic decoration, meat inspection, chain stores
1000	Visual tasks very difficult, i.e. details to be seen are very small (2–3 minutes of arc) and can be of very low contrast. Also accurate colour judgements may be required	General inspection, electronic assembly, gauge and tool rooms, retouching paintwork, cabinet making, supermarkets
1500	Visual tasks extremely difficult, i.e. details to be seen extremely small (1–2 minutes of arc) and of low contrast. Visual aids and local lighting may be of advantage	Fine work and inspection, hand tailoring, precision assembly
2000	Visual tasks exceptionally difficult, i.e. details to be seen exceptionally small (< 1 minute of arc) with very low contrasts. Visual aids and local lighting will be of advantage	Assembly of minute mechanisms, finished fabric inspection

Source: Chartered Institution of Building Services Engineers, CIBSE 1994 Interior Lighting Code.

Table 7.2   Flow chart for obtaining the design maintained illuminance from the standard maintained illuminance

Standard maintained illuminance (lx)	Task size and contrast — Are the task details unusually difficult to see?	Are the task details unusually easy to see?	Task duration — Is the task done for an unusually long time?	Is the task done for an unusually short time?	Error risk — Do errors have unusually serious consequences to people, plant or product?	Design maintained illuminance (lx)
200	Yes — 200	200 — Yes	200	200 — Yes	200 — 200	
	250	250 — Yes	250	250 — Yes	250 — 250	
300	Yes — 300 — Yes	300 — Yes	300 — Yes	300 — Yes	300 — 300	
	400	400 — Yes	400 — Yes	400 — Yes	400 — 400	
500	Yes — 500 — Yes	500 — Yes	500 — Yes	500 — Yes	500 — 500	
	600	600 — Yes	600 — Yes	600 — Yes	600 — 600	
750	Yes — 750 — Yes	750 — Yes	750 — Yes	750 — Yes	750 — 750	
	900	900 — Yes	900 — Yes	900 — Yes	900 — 900	
	1000 — Yes	1000 — Yes	1000 — Yes	1000 — Yes	1000 — 1000	
			1300	1300 — 1300		
			1500	1500 — 1500		

Note: The intermediate values (250, 400, 600 and 900 lx) are a compromise when it is not possible to move a complete step in the scale.
Source: Chartered Institution of Building Services Engineers, CIBSE 1994 Interior Lighting Code.

■ Design maintained illuminance (DMI) – application of Table 7.2 to the RMI, e.g. if the library was mainly for research, then it might be reasonable to raise the RMI from 300 to 500 lx.

Designing a general layout

The first stage is to decide on the type of luminaire to be used. This decision may seem premature, but, unless using computer programs, it is difficult to achieve anything without immediate access to specific photometric data. Alternative approaches can be used where all the design requirements are established, and luminaires are then found to fit these requirements. This method is given in CIBSE Technical Report No. 15: *Multiple criterion design*. However, it is easier to explain the design process by the first method.

Next, the total number of lamp lumens is calculated to achieve the recommended illuminance, and finally the layout is planned.

Determination of lumens required

A commonly accepted method involves the use of the 'lumen formula', which is

$$\text{Initial lamp lumens required} = \frac{E \times A}{\text{UF} \times \text{MF}} \qquad [7.1]$$

where E in the design maintained illuminance (lx), A the area of working plane (m²), UF the utilisation factor and MF the maintenance factor.

Utilisation factor

This is a measure of the efficiency of the lighting scheme and is the proportion of lamp flux that reaches the working plane. Some will come direct and some will come after reflection from other room surfaces.

Table 7.3 shows a typical set of data, and to determine the precise value involves details of the room dimensions and surface reflectances.

The room dimensions are required in terms of the room index, where

$$\text{Room index} = \frac{\text{length} \times \text{width}}{(\text{length} + \text{width}) \times \text{height of lamp above working plane}} \qquad [7.2]$$

Maintenance factor

When a scheme is installed, the illuminance will decline from day 1. There are four measurable reasons which are represented in the expression

$$\text{MF} = \text{LLMF} \times \text{LSF} \times \text{LMF} \times \text{RSMF} \qquad [7.3]$$

where LLMF is the lamp lumen maintenance factor, LSF is the lamp survival factor (used only for group replacement programmes), LMF is the luminaire maintenance factor and RSMF is the room surface maintenance factor. LLMF and LSF vary with the type of lamp being used and LMF and RSMF vary with the type of luminaire, the location and the frequency of cleaning.

In the past, lighting was often designed for a set of average conditions. This could introduced large errors into the final results and the present aim is to provide a more accurate prediction of the following:

■ The initial illuminance when the lamps are new and the luminaires and room surfaces clean.
■ A maintained illuminance below which the actual illuminance should not fall. This value must be based on a programme of replacement and cleaning.

To encourage designers and manufacturers to use the method, CIBSE has produced some generalised data to obtain LLMF and LSF (Table 7.4). This is an interim step and the hope is that manufacturers will produce their own tables.

Tables 7.4, 7.5 and 7.6 are an abbreviated form of the data given in the CIBSE 1994 Lighting Code. There is sufficient data to show how MF can be determined, but the full data have to be obtained from the Code.

Table 7.3   Photometric data for Thorn prismatic enclosure-type luminaire to be used with a range of fluorescent lamps

Description	Single prismatic controller 1800 mm
Report number	500/IL/5295/I
Light Output Ratio	Up 0.26 Down 0.53 Total 0.79
Max. Spacing to Height Ratio (SHR MAX)	1.92

Luminous intensity (cd/1000 lm)			Aspect factors		
Angle (°)	Transverse plane (T)	Axial plane (A)	Angle (°)	Parallel plane	Perpendicular plane
0	132	132	0	0.000	0.000
5	132	131	5	0.087	0.004
10	139	130	10	0.173	0.015
15	147	126	15	0.256	0.033
20	155	122	20	0.334	0.058
25	160	116	25	0.407	0.088
30	160	109	30	0.473	0.123
35	155	102	35	0.532	0.160
40	145	93	40	0.584	0.200
45	132	83	45	0.627	0.239
50	118	71	50	0.661	0.277
55	105	57	55	0.687	0.311
60	96	39	60	0.704	0.338
65	89	23	65	0.714	0.356
70	90	12	70	0.718	0.367
75	94	6	75	0.720	0.373
80	99	1	80	0.721	0.375
85	101	0	85	0.721	0.375
90	102	0	90	0.721	0.375
95	101	0			
100	96	0			
105	93	0			
110	94	0			
115	96	0			
120	95	0			
125	89	0	Luminance distribution (cd/m²/1000 lm)		
130	82	0	Angle (°)	Transverse plane (T)	Axial plane (A)
135	72	0	45	553.3	606.9
140	63	0	50	502.9	571.1
145	54	0	55	458.7	513.8
150	44	0	60	433.6	403.3
155	34	0	65	419.3	281.4
160	22	0	70	446.7	181.4
165	10	0	75	497.0	119.9
170	5	0	80	564.4	29.8
175	0	0	85	630.0	0.0
180	0	0			

The "Luminance distribution (cd/m²/1000 lm)" header and its sub-header columns (Angle (°), Transverse plane (T), Axial plane (A)) together with the data rows 45–85 form a separate sub-table on the right side, aligned with Transverse angle rows 125–180.

Table 7.3 cont'd

Utilisation factors UF												SHR NOM = 1.50
Room reflectances			Room index									
C	W	F	0.75	1.00	1.25	1.50	2.00	2.50	3.00	4.00	5.00
0.70	0.50	0.20	0.41	0.47	0.52	0.55	0.60	0.63	0.66	0.69	0.71
	0.30		0.36	0.42	0.47	0.50	0.56	0.59	0.62	0.66	0.68
	0.10		0.32	0.38	0.43	0.47	0.52	0.56	0.59	0.63	0.66
0.50	0.50	0.20	0.37	0.42	0.46	0.49	0.53	0.55	0.57	0.60	0.61
	0.30		0.33	0.38	0.42	0.45	0.49	0.52	0.55	0.57	0.59
	0.10		0.29	0.34	0.39	0.42	0.47	0.50	0.52	0.56	0.58
0.30	0.50	0.20	0.33	0.37	0.40	0.43	0.46	0.48	0.49	0.51	0.53
	0.30		0.29	0.34	0.37	0.40	0.43	0.46	0.48	0.50	0.51
	0.10		0.27	0.31	0.35	0.38	0.41	0.44	0.46	0.48	0.50
0.00	0.00	0.00	0.23	0.26	0.28	0.30	0.33	0.35	0.36	0.38	0.39

CIE Flux Code 52/80/92/58/79.
Flux fraction ratio 0.72.

Conversion terms					
Luminaire length (mm)	600	1200	1500	1800	2400
Wattage (W)	1×18	1×36	1×65	1×70	1×100
Conversion factors (PC and UF)	1.03	1.09	1.03	1.00	1.05

Source: Thorn Lighting Ltd.

Table 7.4 Lamp lumen maintenance factor (LLMF) and lamp survival factor (LFS) for a selection of lamps

Lamp type		Use (h)		
		6000	10 000	12 000
Fluorescent	LLMF	0.87	0.85	0.84
(Triphosphor)[a]	LSF	0.99	0.96	0.95
Metal halide	LLMF	0.72	0.66	0.63
	LSF	0.91	0.83	0.77
Sodium	LLMF	0.91	0.88	0.87
High pressure	LSF	0.96	0.92	0.89

[a] Using electronic control.
Note: LLMF is the proportion of initial lamp lumens. LSF is the proportion of lamps surviving. If lamps are replaced as soon as they fail then LSF will always be 1.

Example of calculating MF

An office lighting scheme comprises 150 W SON lamps in uplighters. The lamps are bulk changed after 10 000 h use, the uplighters are cleaned every 6 months and the room surfaces every year. The office is large open plan. From Table 7.4, LLMF

Table 7.5 Luminaire maintenance factor (LMF) for a selection of luminaire types and environments

| | Cleaning interval (years) | | | | | | | | |
| | Environment | | | Environment | | | Environment | | |
Luminaire type	C	N	D	C	N	D	C	N	D
Batten	0.95	0.92	0.88	0.93	0.98	0.83	0.91	0.87	0.80
Enclosed IP2X	0.92	0.87	0.83	0.88	0.82	0.77	0.85	0.79	n/r[a]
Uplight	0.92	0.89	0.85	0.86	0.81	n/r[a]	0.81	n/r[a]	n/r[a]

[a] n/r = not recommended.
C, clean atmosphere; N, normal; D, dirty.

Table 7.6 Room surface maintenance factor (RSMF)

| | | Cleaning interval (years) | | | | | |
| | Luminaire flux | Environment | | | Environment | | |
Room index	distribution[a]	C	N	D	C	N	D
	Direct	0.98	0.96	0.95	0.96	0.95	0.94
2.5–5.0	General	0.92	0.88	0.85	0.89	0.85	0.81
	Indirect	0.88	0.82	0.77	0.84	0.77	0.70

[a] See Table 2.2.

is 0.88 and LSF is 0.92 (no spot replacement). From Table 7.5, LMF is 0.92 (clean environment). From Table 7.6, RSMF is 0.88. Therefore,

$$MF = 0.88 \times 0.92 \times 0.92 \times 0.88$$
$$= 0.66$$

If the MF comes below 0.65 it could indicate too long a cleaning interval being used, or lamps being used for too long, or the wrong type of luminaire.

Planning the layout

The spacing of the luminaires should be such that the ratio of the minimum to the average illuminance over an unobstructed task area should be not less than 0.8. There are also recommendations on the ratios of illuminances on the working plane, wall and ceilings which will be discussed later in the chapter.

The required uniformity should be achieved if the layout conforms to the manufacturer's quoted maximum spacing to height ratio (S/h_m). The values of S and h_m are illustrated in Fig. 7.1.

If linear luminaires are used in continuous rows, then the manufacturer may quote a maximum transverse spacing to height ratio. This indicates the maximum spacing between parallel rows of luminaires.

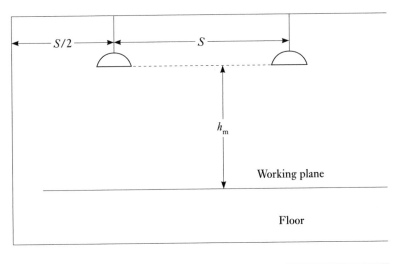

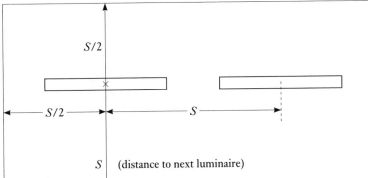

Fig. 7.1 Spacing to mounting height ratio

Example

A lighting scheme is required for a small drawing office, 6 m square, with a 3 m ceiling height. Luminaires are to be used whose photometric data are given in Table 7.3.

The office is clean and the room reflectances are 0.7 ceiling, 0.3 walls and 0.2 floor.

Solution

Step 1 Select the mounting height. In this case the ceiling is relatively low and the luminaires would be mounted direct to the ceiling.

$$\text{Mounting height } h_{\mathrm{m}} = \text{ceiling height} - \text{desk height}$$
$$= 3.0 - 0.8$$
$$= 2.2 \text{ m}$$

Step 2 Determine the illuminance and, hence, the lumens required. The illuminance would normally be selected from the CIBSE schedule. However, a drawing office is an example given in Table 7.1 and the value is 750 lx. Using eqn [7.1], values of UF and LLF must be determined. The room index (eqn [7.2]) is given by

$$RI = \frac{6 \times 6}{12 \times 2.2}$$

$$= 1.4$$

From Table 7.3,

UF = 0.49

To calculate the MF some decisions are needed:

■ Triphosphor fluorescent lamps will be used.
■ They will be bulk replaced, with spot replacement of any individual failures, at 12 000 h intervals of use.
■ The luminaires have a general diffusing distribution.
■ The room surfaces and luminaires are cleaned each year.

Therefore,

$$MF = 0.84 \times 1.0 \times 0.95 \times 0.92$$
$$= 0.73$$

$$\therefore \quad \text{Lamp lumens} = \frac{750 \times 6 \times 6}{0.49 \times 0.73}$$

$$= 75\ 472\ \text{lm}$$

Step 3   Determine the minimum number of lighting points. This may not produce a satisfactory layout but the golden rule is that this number can be increased but not reduced.

The luminaire has a quoted maximum S/h_{m} ratio of 1.92. If $h_{\mathrm{m}} = 2.2$ m, then

$$S = 2.2 \times 1.92$$
$$= 4.2\ \text{m}$$

So the minimum number of points is 2×2, i.e. 4.

Step 4   Can this number achieve 750 lx with sensible lamping? The available lamp sizes are shown in Table 7.3 and range from a 18 W to a 100 W fluorescent lamp. (A range of typical lumen outputs is given in Appendix C.)

For a layout of four points:

$$\text{Lumens per point} = \frac{75\ 472\ \text{lm}}{4}$$

$$= 18\ 868\ \text{lm}$$

No single lamp approaches this initial lumen value. The nearest is the 100 W triphosphor tube with 9400 lm. A possible solution is to use twin-lamp luminaires, but a new set of calculations is required. As it is a drawing office it would be more advantageous to increase the number of points to reduce the risk of shadows. A suggested layout is shown in Fig. 7.2.

An alternative approach could be to provide 500 lx general lighting and 250 lx extra on the boards using local lighting.

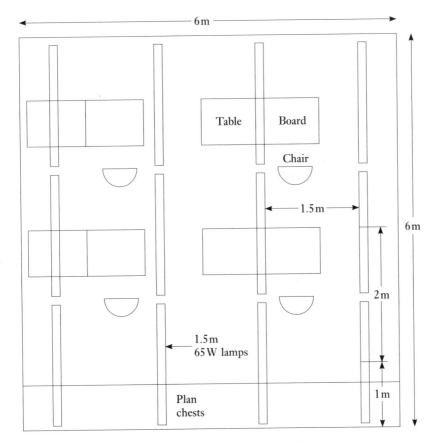

Fig. 7.2 Lighting layout in a drawing office

Types of layouts

General lighting

The simplest and often the most effective layout is the one already described. A regular array of luminaires is planned to achieve an overall level of illuminance. This provides sufficient light for the specified visual task to be carried out anywhere within the room. The pattern of lights can be varied, but in general it is better to keep the lines of luminaires parallel to the room major axis. Diagonal and 'herring bone' patterns can cause a disturbing visual effect.

There are objections to this type of layout, three of which are:

1. The lighting is monotonous and flat. This normally refers to the modelling or degree of shadow, and can be related to the ratio of the vertical and horizontal illuminances. The amount of shadow is, to some extent, a function of the light

distribution, and luminaires with a greater degree of downward light produce deeper shadows. A general diffusing luminaire will produce a much flatter lighting effect.

2. The lighting has no sense of flow. In daytime, with side windows, there is a sense of direction of light from the windows. This is absent at night, although it could be achieved by using angled luminaires. It is suggested that:
   (a) the flow of light in any one part of an interior should be more noticeable in one direction than in any other;
   (b) where possible, the dominant direction of flow should be at 30° or more from the vertical, and the direction of flow may be varied in different parts of the space.

   Achievement of these aims is not difficult when combining daylight and electric light, but a very sophisticated lighting scheme is needed to achieve it using electric lighting only.

3. The layout is extravagant. This implies that there is more light than is necessary in certain parts of the room. A minimum design level for normal working interiors is 200 lx, and if the task requires 750 lx, but is performed in only certain parts of the interior, then it is reasonable to provide an overall layout giving 200 lx with localised additional lighting to provide 750 lx. This is best achieved if the installation is flexible, e.g. it is mounted on an electrical trunking system so that luminaires can be moved if the working area is changed.

Local lighting

There are certain visual tasks which can never be adequately lit by general lighting. The obstructions of the machine may obscure the lighting, or visibility may depend on the revealing of texture by shadow which requires the use of an adjustable small source of light, or a very high local illumination may be needed.

When local lights are used, the lamps must be shaded both from the view of the operator and from people in adjacent work areas.

Localised lighting

Where the furniture or workspace is fixed and is likely to remain so, the lighting can be related. An example is a library which uses localised lighting between the book stacks with, perhaps, local lighting on the reading tables and general lighting in the main circulation area.

The advantages are an improved appearance and more effective lighting. A disadvantage is the lack of flexibility in the use of the room.

Scalar illuminance

Recommended lighting levels are normally in terms of horizontal illuminance. It is arguable that the restriction to the horizontal plane is unrealistic, as the visual field is in all planes and lighting from the side, such as from windows, may give a comparatively low illuminance on horizontal planes but a much higher value on vertical planes.

The scalar illuminance is the amount of flux falling on a sphere of unit surface area at a point in space. The unit is the lux. In the simplest case, if light were flowing in a downward vertical direction, then E_h at a point would be the incident lumens per unit horizontal area.

If this area were a disc of radius r metres, then

$$E_h = \frac{\text{flux}}{\pi r^2} \text{ lx}$$

The spherical or scalar illuminance at the same point would be

$$E_s = \frac{\text{flux}}{\text{area of sphere radius } r}$$

$$= \frac{\text{flux}}{4\pi r^2}$$

or

$$E_h = 4E_s \qquad\qquad [7.4]$$

These two quantities are illustrated in Fig. 7.3.

Calculation of scalar illuminance in an interior

Scalar values are specified for areas such as corridors. The calculation is a fairly simple extension of that required to calculate the horizontal illuminances (E_h). The value is given by

$$E_s = E_h(K + 0.5\rho) \text{ lx} \qquad\qquad [7.5]$$

where K is a value depending on the intensity distribution and room reflections and ρ is the average reflectance of the floor cavity. Figure 7.4 gives values of K for medium room reflectance and some typical luminaires.

Example

A room has an average horizontal illuminance of 500 lx with wall reflectance 0.3 and floor cavity reflectance 0.2. The luminaires are recessed diffusing panels and the room index is 2.0. What is the scalar illuminance?

Solution

From Fig. 7.4 the value of K is 0.36. Therefore,

$$E = 500(0.36 + (0.5 \times 0.2)) \text{ lx}$$
$$= 230 \text{ lx}$$

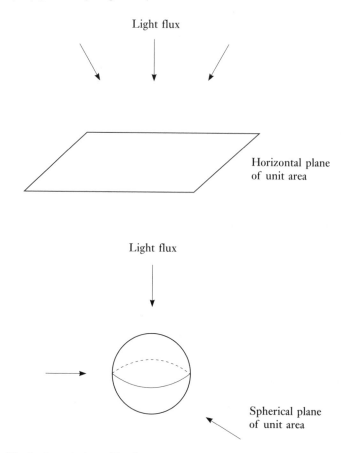

Fig. 7.3   The horizontal planar illuminance (E_h) and scalar illuminance (E_s)

Calculation of cylindrical illuminance in an interior

Cylindrical illuminance can give a better guide to the lighting of vertical surfaces. Examples are storage racks, library shelves and supermarkets.

As with scalar illuminance, a general formula has been evolved which is sufficiently accurate for most purposes:

$$E_c = 1.5E_s - 0.25E_h (1 + RE_f)$$ [7.6]

K is obtainable from Fig. 7.4 and RE_f is the reflectance of the floor cavity.

Relative luminance of room surfaces

Calculations of illuminance on the working plane ensure that there will be sufficient light for the task. They do not indicate the general luminance pattern in the room. To find this, the relative illuminances on the working planes, walls and ceiling must

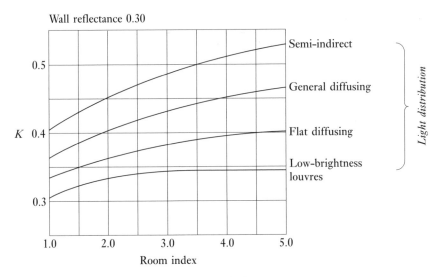

Fig. 7.4 Value of *K* for a variety of typical luminaires

be calculated. Combining these with a knowledge of the surface reflectances enables limited guidance to be made on an acceptable range of luminances. These are illustrated in Fig. 7.5.

The figure displays a wide range, and while it is possible to calculate relative illuminances to a fair degree of accuracy, simpler, less precise methods are more likely to be used.

Table 7.7 from the CIBSE 1984 Lighting Code was provided to act as an aid in the selection of a lighting system. It included suggested patterns of room brightness in a diagram whose explanation is given in Table 7.8. This indicates the anticipated relative brightness of the ceiling and walls in a room with average surface reflectances for that specific type of luminaire. These tables are not in the 1994 Code.

Approximate method of calculating room illuminance ratios

The following data are required:

- The reflectance and size of room surfaces.
- The ULOR (Upward Light Output Ratio).
- The $UF_{0,0,0}$, which is the utilisation factor for zero room reflectances.
- The average horizontal illuminance on the working plane (E_h).
- The initial total lamp lumens (F).

The luminance of room surfaces comprises

[Direct illuminance (from luminaires) + reflected illuminance (from other surfaces)] × reflectance of surface

The direct illuminance on the working plane (E_{dwp}) is given by the formula

Table 7.7 Typical luminaire characteristics (excerpt)

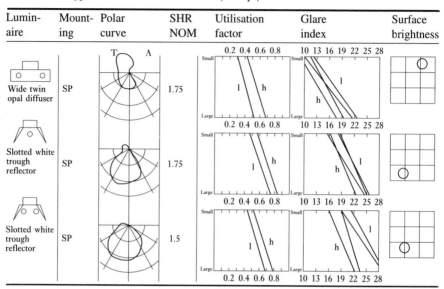

Lumin-aire	Mount-ing	Polar curve	SHR NOM	Utilisation factor	Glare index	Surface brightness
Wide twin opal diffuser	SP		1.75			
Slotted white trough reflector	SP		1.75			
Slotted white trough reflector	SP		1.5			

Source: Chartered Institution of Building Services Engineers, CIBSE 1984 Lighting Code.

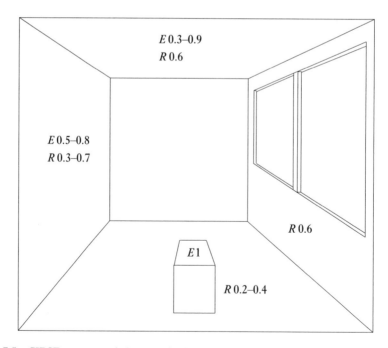

Fig. 7.5 CIBSE-recommended range of reflectances and illuminances for room surfaces of a working interior (courtesy Chartered Institution of Building Services Engineers. CIBSE 1994 Lighting Code)

Table 7.8 Key to pattern of room surface in brightness in Table 7.6

Bright ceiling		Bright ceiling		Bright ceiling	
	Dull wall		Medium wall		Bright wall
Medium ceiling		Medium ceiling		Medium ceiling	
	Dull wall		Medium wall		Bright wall
Dull ceiling		Dull ceiling		Dull ceiling	
	Dull wall		Medium wall		Bright wall

Increasing ceiling brightness →

Increasing wall brightness →

Source: Chartered Institution of Building Services Engineers, CIBSE 1984 Interior Lighting Code.

$$E_{dwp} = \frac{F \times UF_{0,0,0} \times MF}{\text{area of working plane}} \qquad [7.7]$$

The direct illuminance on walls (E_{dw}) will be a function of the downward flux not reaching the working plane directly:

$$E_{dw} = \frac{F \times (UF - UF_{0,0,0}) \times MF}{\text{area of four walls}} \qquad [7.8]$$

For the direct illuminance on the ceilings (E_{dc}) the upward flux will all fall directly onto the ceiling cavity:

$$E_{dc} = \frac{E \times ULOR \times MF}{\text{area of ceiling cavity}} \qquad [7.9]$$

Each room surface now acts as a light source and provides the indirect or reflected illuminance component. The total illuminance on each surface will be the sum of the direct and reflected components.

E_h = horizontal illuminance originally calculated for the lighting scheme

The next step is to calculate the reflected component of illuminance on the walls and ceiling. Before making precise calculations it is relevant to look at the recommendations. These are usually quoted in terms of a range of reflectances and illuminances, since luminance is proportional to illuminance times reflectance.

It is unrealistic to quote precise values as a wide range of luminances is acceptable. Bearing this in mind, a reasonable approximation of the illuminances on walls and ceilings can be obtained by assuming that the reflected component is the same for all surfaces.

For the working plane:

$$E_{wp} = E_{dwp} + E_{reflected}$$
$$\therefore \quad E_{reflected} = E_{wp} - E_{dwp}$$

For the walls:

$$E_w = E_{dw} + E_{reflected}$$

For the ceiling:

$$E_c = E_{dc} + E_{reflected}$$

Example

The average illuminance in an office is 500 lx. The room dimensions are $L = 8$ m, $W = 6$ m, $H = 2.65$ m. The desk height is 0.75 m and h_m is $(2.65 - 0.75)$ m or 1.9 m. The room surfaces have average reflectances

$$\rho_c = 0.7$$
$$\rho_f = 0.2$$
$$\rho_w = 0.5$$

The photometric data are from Table 7.3. First, the lumens needed for a maintained illuminance of 500 lx, using the maintenance schedule as in the previous example, are given by

$$F = \frac{500 \times (8 \times 6)}{UF \times MF} \text{ lm}$$

The room index is given by

$$RI = \frac{8 \times 6}{(8 + 6) \times 1.9} = 1.8$$
$$\therefore \quad UF = 0.58$$

$MF = 0.73$ as in the previous example (p. 156). Hence

$$F = \frac{500 \times 48}{0.58 \times 0.73}$$
$$= 56\,683 \text{ lm (initial)}$$
$$E_{dwp} = \frac{56\,683 \times UF_{0,0,0} \times MF}{8 \times 6}$$
$$UF_{0,0,0} = 0.32$$
$$\therefore \quad E_{dwp} = \frac{56\,683 \times 0.32 \times 0.73}{48}$$
$$= 276 \text{ lx}$$
$$E_{reflected} = 500 - 276$$
$$= 224 \text{ lx}$$

It is interesting to note that nearly half the average horizontal illuminance is by reflection off the room surfaces.

From eqn [7.8]

$$E_{dw} = \frac{56\,683 \times (0.53 - 0.32) \times 0.73}{1.9 \times 2 \times (8 + 6)}$$

and so, assuming luminaires are mounted direct to the ceiling,

$$E_{dw} = 163 \text{ lx}$$

From eqn [7.9]

$$E_{dc} = \frac{56\,683 \times 0.26 \times 0.73}{8 \times 6}$$

$$= 224 \text{ lx}$$

Adding the reflected light:

$$E_w = 163 + 224 = 387 \text{ lx}$$
$$E_c = 224 + 224 = 448 \text{ lx}$$

Figure 7.5 shows suggested illuminance ratios. For the walls the range is 0.5–0.8. In this example the ratio is 387/500 or 0.77.

For the ceiling the range is 0.3–0.9. In this example the ratio is 448/500 or 0.89. Both these ratios fall within the acceptable ranges.

The luminance values are given by

$$\text{Luminance} = \frac{\text{illuminance} \times \text{reflectance}}{\pi} \text{ cd/m}^2 \qquad [7.10]$$

Therefore, for the working plane:

$$\text{Luminance} = \frac{500 \times 0.2}{\pi} = 32 \text{ cd/m}^2$$

For the walls:

$$\text{Luminance} = \frac{387 \times 0.5}{\pi} = 62 \text{ cd/m}^2$$

For the ceiling:

$$\text{Luminance} = \frac{448 \times 0.7}{\pi} = 100 \text{ cd/m}^2$$

It must be emphasised that this is an approximate method devised by the author. More precise methods can be found in CIBSE Technical Memorandum No. 5 and Technical Report No. 15, using transfer functions.

Calculation of glare

In both daylight and electric light there is always a need to restrict the luminances within the normal field of view, otherwise an unacceptable degree of glare will be experienced.

Disability glare

This can be defined as glare which impairs the ability to see detail without necessarily causing visual discomfort. As it is a real reduction in the ability to see, it can be measured in terms of visual performance, and its effect can be expressed as a shift in the adaptation level of the eye (see Chapter 1). If the adaptation level is raised by a light source, such as an unshaded lamp or window, then the eye becomes less sensitive to small differences in brightness.

One formula for expressing this change in adaptation level is

$$\text{Adaptation luminance} = L_o + \frac{kE}{\theta^2} \qquad [7.11]$$

where L_o is the original average luminance of the field of view, E is the illuminance from the glare source at the eye, θ is the angle between the line of sight and the direction of the glare source and k is a constant related to the age of the observer.

Present attempts to identify and design for the avoidance of disability glare in the task are based on the work on contrast rendition factors (see Chapter 1). Details are contained in the CIE Publication No. 19: *A unified framework of methods for evaluating visual performance aspects of lighting*. There are subsequent papers in the UK, principally by the Building Research Establishment and the Electricity Council, but the stage has not yet been reached when specific recommendations can be made and accepted. However, reduction in contrast sensitivity is used as a measure of glare from road lighting lanterns and this is discussed further in Chapter 10. Luminance limits are used in hospitals and offices.

Discomfort glare

This can be defined as: *Glare which causes visual discomfort without necessarily impairing the vision*. The degree of discomfort will depend on the type of location and angle of view; for instance, people will tolerate a much brighter installation in a supermarket, where they are continually on the move and looking in all directions, than in a classroom, where the direction of view is fixed, the visual task more exacting and the occupants have more time to become aware of the lighting. This is very much a subjective assessment; it cannot be measured in terms of performance, but only in relative degrees of discomfort. Work at the Building Research Establishment has provided the basic data for the CIBSE system of classifying discomfort glare, known as the *glare index system*. Glare from a light source can be expressed as

$$\text{Discomfort glare } (g) = \frac{L_s^{1.6} \times \omega^{0.8}}{L_B \times P^{1.6}} \times 0.45 \qquad [7.12]$$

where L_s is the luminance of the source, L_B is the average luminance of the background, ω is the angular size of the source and P is the position index which indicates the effect of the position of the source on glare.

Table 7.9 CIBSE recommended limiting glare indices for different classes of visual task

Class of visual task	Examples	Limiting glare index
Critical, often with a fixed direction of view	Drawing offices, very fine inspection	16
Critical, but general direction of view	Offices, libraries, computer buildings	19
Ordinary task with general direction of view	Kitchens, reception areas, fine assembly work	22
Large task or limited viewing time	Stock rooms, medium assembly work	25
No specific visual task or direction of view	Rough industrial work, indoor parking areas	28

To calculate the glare from each light point for a range of room positions would be too time consuming. The method can be simplified by considering a number of standard conditions and making adjustments as necessary. The standard conditions and assumptions are:

- Glare is additive; therefore, for a complete installation, the glare from all the luminaires would be the same as for a single luminaire of the same luminance L_s but whose angular size ω is the total of all the individual luminaires.
- The direction of view is horizontal, straight across the room and 1.2 m above floor level.
- The room is rectangular and has a regular array of luminaires.

The calculations and corrections depend on the tabular data being used. The broad steps are:

1. Find the initial glare index for the particular type of luminaire in a specific room under standard conditions. The index is 10 log g, where g, the discomfort glare, can be found from [7.12].
2. Apply various correction factors as required by the system being used; this will give the final glare index. This value should not exceed the limiting glare index recommended for that particular type of location. A list of typical limiting glare indices is given in Table 7.9.

The full method is given in CIBSE Technical Memorandum No. 10. It is fairly laborious and some manufacturers now produce glare data for specific luminaires.

Table 7.10 shows a typical glare data sheet which, in this case, relates to the luminaire data given in Table 7.3. The glare data sheet is based on a number of assumptions:

1. The luminaire is 2 m above eye level.
2. The lamp emits 1000 lumens.
3. The spacing to height ratio is 1:1.

Table 7.10   Glare data for Thorn prismatic enclosure-type luminaire (see Table 7.3 for general photometric data)

Glare indices										
Ceiling reflectance	0.70	0.70	0.50	0.50	0.30	0.70	0.70	0.50	0.50	0.30
Wall reflectance	0.50	0.30	0.50	0.30	0.30	0.50	0.30	0.50	0.30	0.30
Floor reflectance	0.14	0.14	0.14	0.14	0.14	0.14	0.14	0.14	0.14	0.14

Room dimension											
X	Y	Viewed crosswise					Viewed endwise				
$2H$	$2H$	7.3	8.5	8.4	9.7	11.1	6.7	8.0	7.8	9.1	10.6
	$3H$	9.4	10.5	10.5	11.6	13.1	8.4	9.5	9.5	10.7	12.2
	$4H$	10.4	11.4	11.5	12.6	14.1	9.1	10.2	10.3	11.4	12.9
	$6H$	11.5	12.5	12.6	13.6	15.2	9.8	10.7	10.9	11.9	13.5
	$8H$	12.1	13.0	13.3	14.2	15.8	10.0	10.9	11.1	12.1	13.7
	$12H$	12.8	13.7	13.9	14.9	16.4	10.1	11.0	11.2	12.2	13.7
$4H$	$2H$	8.1	9.1	9.2	10.3	11.8	7.6	8.7	8.8	9.8	11.4
	$3H$	10.5	11.4	11.6	12.6	14.1	9.6	10.5	10.8	11.7	13.2
	$4H$	11.7	12.5	12.8	13.7	15.3	10.5	11.3	11.7	12.5	14.1
	$6H$	13.0	13.8	14.2	15.0	16.6	11.4	12.1	12.6	13.3	14.9
	$8H$	13.8	14.5	15.0	15.7	17.3	11.7	12.4	12.9	13.6	15.2
	$12H$	14.6	15.2	15.8	16.4	18.1	11.9	12.5	13.1	13.8	15.4
$8H$	$4H$	12.3	13.0	13.5	14.2	15.8	11.3	12.0	12.5	13.2	14.8
	$6H$	13.9	14.5	15.2	15.7	17.4	12.5	13.1	13.7	14.3	15.9
	$8H$	14.9	15.4	16.1	16.6	18.2	13.0	13.5	14.2	14.7	16.4
	$12H$	15.9	16.4	17.2	17.6	19.3	13.4	13.8	14.6	15.1	16.7
$12H$	$4H$	12.4	13.0	13.6	14.2	15.8	11.6	12.2	12.8	13.4	15.0
	$6H$	14.1	14.7	15.4	15.9	17.5	12.9	13.4	14.1	14.6	16.2
	$8H$	15.2	15.7	16.5	16.9	18.6	13.5	14.0	14.7	15.2	16.9
	$12H$	16.1	16.5	17.4	17.8	19.4	13.7	14.1	15.0	15.4	17.0

Glare index conversion:					
Luminaire length (mm)	600	1200	1500	1800	2400
Conversion	+3.82	+1.41	+0.63	0	−1.0

Source: Thorn Lighting Ltd.

Table 7.11 gives corrections for factors 1 and 2. No corrections are available for 3.

Initial glare indices are tabulated according to room dimensions and reflectances. Figure 7.6 shows the method of specifying the room dimensions. The Y dimension is always parallel to the line of sight and the X dimension is across the line of sight. They are both expressed as multiples of the mounting height above eye level.

One view of the room will show the ends of the luminaires (endwise view) and the other view will show the sides (crosswise view).

When the initial glare index has been found it must be corrected for:

Table 7.11 Correction data for use with Table 7.10

Initial lamp lumens	Conversion term	Height H above 1.2 m eye level (m)	Conversion term
100	−6.0	1	−1.2
150	−4.9	1.5	−0.5
200	−4.2		
300	−3.1	2	0.0
500	−1.8		
700	−0.9	2.5	+0.4
1 000	0.0	3	+0.7
1 500	+1.1	3.5	+1.0
2 000	+1.8		
3 000	+2.9	4	+1.2
5 000	+4.2	5	+1.6
7 000	+5.1	6	+1.9
10 000	+6.0		
15 000	+7.1	8	+2.4
20 000	+7.8	10	+2.8
30 000	+8.9	12	+3.1
50 000	+10.2		

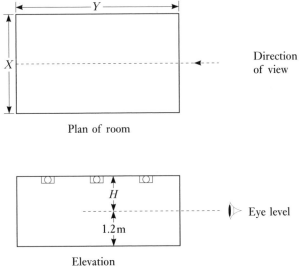

Plan of room

Elevation

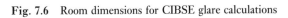

Fig. 7.6 Room dimensions for CIBSE glare calculations

- Mounting height above 1.2 m eye level if this differs from 3 m.
- Total downward luminous flux if this differs from 1000 lm.
- Extra correction terms if the published glare index table covers a variety of luminaire sizes or lamp types.

These correction terms are added to (or subtracted from) the initial glare index to give the final glare index of the installation.

Example

Referring to the scheme designed earlier in this chapter and illustrated in Fig. 7.3, what is the glare index?

Solution

The room is square, so $X = Y = 6$ m. The ceiling height is 3 m and the height of eye level for a seated person is 1.2 m above the floor.

$$H = 3 - 1.2$$
$$= 1.8 \text{ m}$$
$$\therefore \quad X = 6 \text{ m}$$
$$= \frac{6H}{1.8}$$
$$= 3.3H = Y$$

The reflectances are 0.7 ceiling, 0.3 wall and 0.2 floor. In this case the worst glare situation is when the luminaires are viewed crosswise. From Table 7.10 the initial glare index is found by interpolation:

X	Y			$Y = 3.3H$	$X = 3.3H$
$2H$	$3H$	10.5	10.8		
	$4H$	11.4			
					11.4
$4H$	$3H$	11.4	11.8		
	$4H$	12.5			

Applying corrections from Table 7.11,

Initial lamp lumens = 5100 lm
Correction = +4.2

Height above eye level is 1.8 m. Therefore,

Correction = −0.2

There is a further glare index conversion of +0.63 (see Table 7.3) for the 1.5 m lamp. Therefore,

Final glare index = 11.4 + 4.2 − 0.2 + 0.63
= 16.03

As the limiting glare index for a drawing office is 16, this installation barely meets the glare limit requirements. It does suggest that glare could be a problem, and another type of luminaire should be selected.

Self-assessment task 7.1

How much more glaring (eqn [7.12]) would a 26 mm diameter fluorescent lamp be than a 35 mm diameter lamp of the same length and lumen output?

Discomfort glare from large areas

The glare index system refers to layouts of individual luminaires. In the case of large luminous areas, such as luminous diffusing ceilings, it is more effective to control discomfort glare by limiting the source brightness at normal viewing angles. It is recommended that overall diffusing luminous ceilings are not used where a glare index of less than 19 is specified. Elsewhere, the luminance should not exceed 500 cd/m^2.

CIE Unified Glare Rating (UGR)

International agreement has been reached on a new system (UGR). This is based on a slightly different formula to the British system (eqn [7.12]). There are not serious differences from the British system which CIBSE continues to recommend. This will change eventually but that time has not yet arrived.

CIE luminance curve system

Some UK manufacturers provide data for the CIE method, and, as will be shown, it is relatively simple to use.

The degree of glare is assessed in terms of luminances in the axial and transverse planes between 45° and 85°. Table 7.12 and Fig. 7.7 are used to check if the luminaire chosen will be acceptable in a room of specific size for a specific class of interior. Table 7.13 lists the five classes, with examples.

Luminaires described in this system use the following terms:

■ *Luminous sides* – luminous side panels with a height greater than 30 mm.
■ *Elongated* – when the ratio of the length to width of the plan luminous area is not less than 2:1.

The steps in the calculation are:

Table 7.12 CIE glare system: relation quality class (A–E) to curve letter (*a–h*)

Quality class	Valid for service illuminance *E* (lx)							
A	2000	1000	500	≤300				
B		2000	1000	500	≤300			
C			2000	1000	500	≤300		
D				2000	1000	500	≤300	
E					2000	1000	500	≤300
Curve letter	*a*	*b*	*c*	*d*	*e*	*f*	*g*	*h*

Table 7.13   Quality classes of interiors for CIE glare system

Class	Quality	Example
A	Very high	Drawing office
B	High	General office
C	Medium	Supermarket
D	Low	Toilets
E	Very low	Iron foundry

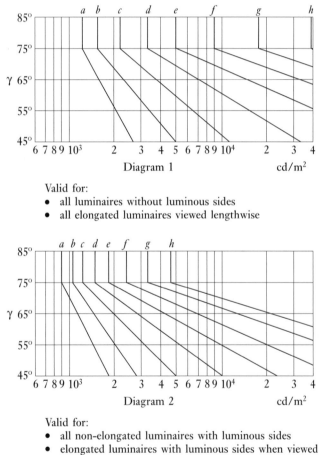

Valid for:
- all luminaires without luminous sides
- all elongated luminaires viewed lengthwise

Valid for:
- all non-elongated luminaires with luminous sides
- elongated luminaires with luminous sides when viewed crosswise

Fig. 7.7   Luminance limitation curves for a stepped scale of glare ratings (0 = no glare, to 6 = intolerable glare) representing quality classes A to E, and for various values of standard service illuminance E (courtesy Philips Lighting Ltd)

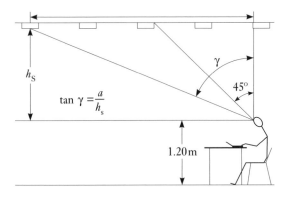

Fig. 7.8 Calculation of critical angle and for glare calculations (courtesy Philips Lighting Ltd)

1. Determine the mean luminances between 45° and 85° of the type of luminaire selected for the installation.
2. Determine the quality, class and illuminance level required for the installation in the new condition.
3. Select the appropriate curve (class and level) of the relevant diagram (Fig. 7.7).
4. Determine the maximum angle to be considered for the room length and height between eye level and the plane of the luminaires. To do this take the horizontal line on the glare limitation diagram for the value of a/h_s thus found. The part of the curve above this line may be ignored. The value of a/h_s is found from Fig. 7.8.
5. Compare the luminance of one luminaire with the selected part of the limiting curve.

Discomfort glare will not result if the value of the luminance given by the limiting curve exceeds the actual luminance of the luminaire over the whole range of emission angles. If the result is otherwise the design will have to be changed; for example, by choosing a different type of luminaire.

Example

Using the previous example and viewing the luminaires crosswise, will the installation comply?

Solution
Luminance data are included in Table 7.3. These values are based on 1000 lm and will need correcting to 5100 lm. As this is a drawing office, Class A would be appropriate. The illuminance level is already designed to 750 lx; the luminaire would be classed as elongated, luminous sided.

From Table 7.12 either curve *b* or *c* could be selected for use in Fig. 7.7. The relevant range of angles for luminance values is from 45° upwards. The upper limit (γ) depends on the room shape, and this is illustrated in Fig. 7.8.

In this example h_s is 1.8 m and a is 6 m; hence

$$\tan \gamma = \frac{a}{h_s}$$

$$= \frac{6}{1.8}$$

$$= 73°$$

So only 45° to 73° need be considered. From Table 7.3 for a transverse plane,

$$\text{Luminance at } 45° = 553 \times \frac{5100}{1000} \text{ cd/m}^2$$

$$= 2820 \text{ cd/m}^2$$

$$\text{Luminance at } 55° = 459 \times \frac{5100}{1000} \text{ cd/m}^2$$

$$= 2340 \text{ cd/m}^2$$

$$\text{Luminance at } 65° = 419 \times \frac{5100}{1000} \text{ cd/m}^2$$

$$= 2136 \text{ cd/m}^2$$

$$\text{Luminance at } 75° = 94 \times \frac{5100}{1000} \text{ cd/m}^2$$

$$= 479 \text{ cd/m}^2$$

Looking at Fig. 7.7, Diagram 2, the 45° value lies on b, the 55° value between b and c and the 65° value between c and d. As with the CIBSE method this is indicating excessive glare for this luminaire in this type of situation.

Alternative methods are used in the USA (visual comfort rating) and in Australia (luminance limit system). Details are given in their respective lighting codes.

UPLIGHTING

Indirect lighting has been hailed as the 'lighting of the 1990s'. It was also the 'lighting of the 1930s' and one may well think 'so what's new?' As in the 1930s, it claims to be glare free. It enables the use of higher lamp wattages than normal with relatively low ceilings, and its revival is closely linked with the computer age. It is still a relatively inefficient method of lighting, highly dependent on a good standard of maintenance and you either like it or you don't.

Self-assessment task 7.2

Can you list any drawbacks to lighting an office to 750 lx using uplighters only?

There are various techniques for planning an uplighter scheme and these notes will deal with two methods and compare their answers:

1. Lumen method using TF tables.
2. Thorn Lighting method – 'equivalent point source'.

Lumen method

Utilisation factors are available or can be calculated for indirect lighting. These can then be used to give an average illuminance value. What this method cannot do is give an indication of uniformity. It just assumes the lighting will be uniform.

To explain the two methods we will apply each one to a set situation.

Example

An office 8 m × 8 m is to be lit by four 250 W MBIF lamps in uplighters. The following information is available:

Ceiling reflectance	0.7
Wall reflectance	0.5
Mounting height	1.8 m (top of uplighter)
Ceiling height	2.8 m
ULOR of uplighter	0.85

Step 1   The ceiling is effectively a cavity, $2.8 - 1.8$ m deep. Because the ceiling cavity, i.e. that upper part of the room which received direct light from the uplighter, is not flat, light will be absorbed by interreflection within the cavity. The effective reflectance of the ceiling cavity will be less than the actual reflectances of the surfaces. To calculate the effective reflectance (R_E) of the ceiling cavity:

$$R_E = \frac{CI \times R_A}{CI + 2(1 - R_A)} \tag{7.13}$$

CI is the cavity index (like room index) and R_A is the average reflectance of the cavity surfaces. Therefore,

$$CI = \frac{\text{area of flat ceiling}}{\text{area of } \frac{1}{2} \text{ walls of cavity}} \tag{7.14}$$

$$= \frac{8 \times 8}{(8 + 8) \times (2.8 - 1.8)}$$

$$= 4$$

$$R_A = \frac{\text{ceiling area} \times \text{reflectance} + \text{cavity wall area} \times \text{reflectance}}{\text{ceiling} + \text{cavity wall areas}} \tag{7.15}$$

$$= \frac{(8 \times 8) \times 0.7 + [2 \times (8 + 8) \times 1.0] \times 0.5}{(8 \times 8) + 2(8 + 8) \times 1.0}$$

$$= 0.63$$

$$R_E = \frac{4 \times 0.63}{4 + 2(1 - 0.63)}$$

$$= 0.53$$

To calculate the room index, RI, take the height from the working plane (0.8 m) to the top of the uplighter (1.8 m):

$$RI = \frac{8 \times 8}{(8 + 8) \times (1.8 - 0.8)}$$

$$= 4$$

Table 7.14   Transfer factor, TF_{CF} (for floor reflectance 20%)

Reflectance		Room index (K)				
R_C	R_W	1.0	1.5	2.0	3.0	5.0
80	50	0.492	0.605	0.674	0.754	0.826
80	30	0.418	0.537	0.613	0.704	0.790
70	50	0.423	0.519	0.578	0.645	0.707
70	30	0.362	0.463	0.528	0.605	0.678
50	50	0.291	0.357	0.396	0.442	0.483
50	30	0.253	0.322	0.366	0.418	0.467

Source: Chartered Institution of Building Services Engineers, CIBSE 1994 Lighting Code.

Step 2   Use of transfer factors   A transfer factor (TF) indicates the proportion of light flux that is transferred from one diffusing surface to another. Part of the transfer is direct and part by interreflection within other surfaces. Thus the TF_{CF} gives the TF from floor to ceiling and vice versa. It will depend on the room shape (room index) and surface reflectances. As already said, the room is considered for uplighting as the space between the top of the uplighter and the working plane and the reflectances are the effective values for the ceiling and floor cavities. The values do not depend on the light intensity distribution of the luminaire.

Table 7.14 shows a set of factors for a range of reflectances and room indices.

Provided the ULOR of the luminaire is known, the utilisation factor for the uplighting scheme is given by

$$UF = ULOR \times TF_{CF} \qquad [7.16]$$

For the example, the effective ceiling cavity reflectance is 0.53 and the ULOR is 0.85. The RI is 4. Thus,

$$TF_{CF} = 0.50$$

from Table 7.14 and by interpolation, and

$$UF = 0.85 \times 0.50$$
$$= 0.425$$

Note that this method only applies for the upward light and although a similar technique is used for downward lighting the equations are more complex.

Assume a maintenance factor of 0.7 (a dangerous thing to do). The initial lumen output of 250 W MBIF is 19 000 lm (from Appendix C). Using the lumen formula,

$$\text{Illuminance} = \frac{4 \times 19\,000 \times 0.425 \times 0.7}{8 \times 8}$$
$$= 353 \text{ lx}$$

Thorn Lighting method

L. Bedocs, J. Lynes and J. Hugill presented a paper to the 1984 National Lighting Conference entitled 'Estimating uplighting'. The main proposal was that if the illuminance from a combination of a single point and large ceiling area was measured on a horizontal working plane, it was possible to work backwards to suggest that

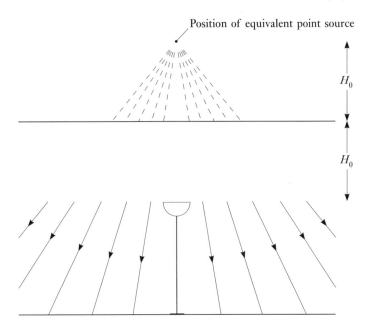

Fig. 7.9 The equivalent point source (courtesy Thorn Lighting Ltd)

there was an equivalent direct luminaire that would give the same distribution as the uplighter and *first* reflection off the ceiling. This is illustrated in Fig. 7.9.

The merit of this proposal is that the illuminance values for different values of H_0 can then be derived. The illuminances due to second and subsequent reflectances are added and this is done with the help of a table.

Photometric data are published for a 2.8 m ceiling height and an example is shown in Fig. 7.10. The graph indicates the horizontal illuminance per 1000 lm at various distances along the working plane.

Ceiling height correction

So far in this explanation the method has only been applied to a ceiling height of 2.8 m and an uplighter height of 1.8 m. Table 7.15 gives corrections that are applied to the lux values and horizontal distances for other heights.

Example

The design stages are:

1. Work out the approximate number of uplighters needed.
2. Calculate the direct illuminances using the illuminance curve.
3. Calculate the interreflected component of illuminance.
4. Finally, check the ceiling luminance to see it does not exceed 1500 cd/m², with an average value not above 500 cd/m².

Description: Freestanding circular cantilever uplight

Conversion Terms

TLL Ref	Catalogue Number	Lamp	LOR
UL018	DUGY 250	250W MBIF/SONDL-E	0.84
UL019	DUSY 150	150W SONDL-E	0.85

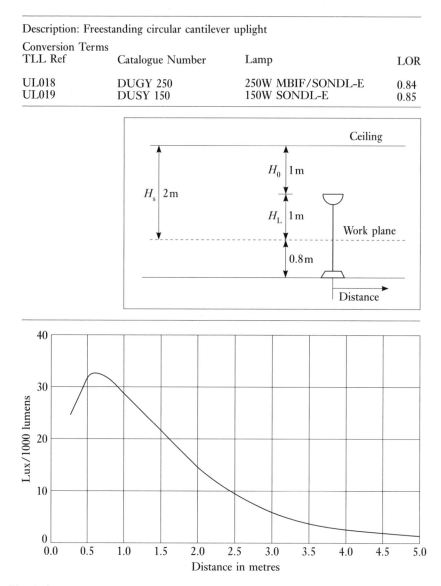

Fig. 7.10   Photometric data for an uplighter (courtesy Thorn Lighting Ltd)

This may all seem long winded but it has the merit that it is based on manufacturers' data.

Step 1 – Lumen calculation   This could be done by the first method but a simple rule of thumb is now provided:

To provide uniform illuminance of 500 lx, a 250 W MBIF lamp covers 15 m^2 and a 250 W SON-DL lamp covers 20 m^2.

In this example there are 4×250 W MBIF lamps/64 m^2, so there is one point per 16 m^2.

Table 7.15 Correction factors for illuminance curve in Fig. 7.10

Ceiling height (m)	Distance (m)	Illuminance (lx)
2.4	0.73	1.86
2.6	0.87	1.33
2.8	1.00	1.00
3.0	1.13	0.78
3.2	1.27	0.62
3.4	1.40	0.51
3.6	1.53	0.43
3.8	1.67	0.36
4.0	1.80	0.31

Step 2 – direct illuminances   An isolux diagram can be drawn for one uplighter. The values must be corrected for actual lamp lumens and an MF applied.

$$\text{Correction factor} = \frac{19\,000 \times 0.7}{1000}$$

$$= 13.3$$

The approximate isolux diagram is shown in Fig. 7.11.

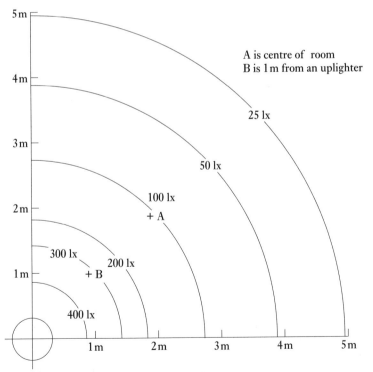

A is centre of room
B is 1 m from an uplighter

Uplighter

Fig. 7.11   Isolux diagram for uplighter (Fig. 7.10): $H_0 = 1$ m, $H_L = 1$ m

Table 7.16 Utilisation factors for second and subsequent reflections

| Room reflectances | | | Room index | | | | | | | | | |
C	W	F	0.75	1.00	1.25	1.50	2.00	2.50	3.00	4.00	5.00	∞
0.70	0.50	0.20	0.125	0.133	0.135	0.136	0.136	0.134	0.132	0.130	0.128	0.114
	0.30		0.064	0.072	0.077	0.080	0.086	0.090	0.092	0.096	0.099	0.114
	0.10		0.021	0.027	0.031	0.036	0.045	0.052	0.057	0.067	0.073	0.114
0.50	0.50	0.20	0.081	0.083	0.084	0.083	0.080	0.078	0.076	0.072	0.069	0.056
	0.30		0.042	0.45	0.048	0.050	0.051	0.052	0.052	0.052	0.053	0.056
	0.10		0.013	0.016	0.019	0.020	0.024	0.028	0.031	0.035	0.037	0.056

With a layout of four uplighters, assuming they are on a spacing of 4 m × 4 m: the illuminance in the room centre, A, on the isolux diagram, is 4 × 105 lx, i.e. 420 lx; at B the illuminance is found by calculating its distance from each of the uplighters

$$E_B = 370 + 2 \times 26 + 0$$
$$= 422 \text{ lx}$$

Step 3 – interreflected illuminance (E_R) Step 2 has calculated the illuminance due to first reflection. Light will continue to be reflected and this requires a lumen formula type of calculation.

For R_E of 0.53 and RI of 4, the UF is 0.081 from Table 7.16. Using the lumen formula,

$$E_R = \frac{4 \times 19\,000 \times 0.081 \times 0.7}{8 \times 8}$$
$$= 67 \text{ lx}$$

Returning to the direct values, Step 2, the ceiling has an actual reflectance of 0.7 and the Thorn data are based on 0.8, so there is a further scaling needed. Adjusted direct values plus interreflected values give

$$E_A = 420 \times \frac{0.7}{0.8} + 67$$
$$= 434 \text{ lx}$$
$$E_B = 422 \times \frac{0.7}{0.8} + 67$$
$$= 436 \text{ lx}$$

It is difficult to derive an average value as these values are only in the central part of the room.

Step 4 – check ceiling luminance   The CIBSE Code recommendation of a limit of 500 cd/m² average and a maximum of 1500 cd/m² applies to the control of discomfort glare. The maximum value can only be checked if the manufacturer provides the intensity distribution of the uplighter.

The easiest check of average luminance is to consider the ceiling as an 'overcast sky'. In this case the sky luminance, in apostilbs, equals the horizontal ground illuminance. For example,

E_{h} of 500 lx $\equiv L_{\mathrm{s}}$ of 500 Asb

$$= L_{\mathrm{s}}\,\frac{500}{\pi}\,\mathrm{cd/m^2}$$

$$= 160\ \mathrm{cd/m^2}$$

The scheme will lie well within 500 cd/m^2 average value.

A more accurate method of calculating luminance at a point on a ceiling is given in Appendix 3 of LG3 1996 using the inverse square law. The drawback is that it requires the upward intensity distribution and this may not be available in catalogues.

EMERGENCY LIGHTING

It is an increasing requirement that commercial, industrial and public buildings are provided with some form of emergency lighting. Requirements vary for different types of buildings and even their geographical location. General guidance is given in BS 5266: 1988 *The emergency lighting of premises.*

However, note that UK standards are in the process of being harmonised with European Standard EN 1838 and, where recommendations may differ, the European figures are given in this chapter as they are likely to be the standards in the future. There is general agreement in principle between BS 5266 and EN 1838 but the lighting levels in EN 1838 are generally higher. A useful guidance booklet published by ICEL (Industry Committee for Emergency Lighting) is available from 207 Balham High Road, London SW17 7BQ.

The following is a summary of the design requirements. The main classifications are:

■ Emergency lighting – lighting provided for use when the normal lighting fails.
■ Escape lighting – that part of emergency lighting which is provided to ensure that the means of escape can be safely and effectively used at all material times.
■ Standby lighting – that part of emergency lighting which may be provided to enable normal activities to continue when the mains fails.

The basic recommendations are that, during the period of use, the horizontal illuminance measured on the centre line of any escape route should never fall below 1.0 lx and in halls and corridors it should be measured at floor level. The emergency system should be in operation within 5 s of the failure of the normal lighting installation. An escape route is taken as a strip 2 m wide.

Care should be taken to avoid abrupt alternations of excessive dark and light areas on the floor.

There are points in these recommendations that require some elaboration:

■ The illumination level quoted, of 1.0 lx, is roughly equivalent to a moonlit night. It is the intention that this should be an absolute minimum, i.e. with an aged lamp and battery, soiled luminaire and at the end of the period of duration. Such a situation would suggest a desirable illumination level of between one-tenth and one-fiftieth of the normal lighting level with a minimum value of 1.0 lx.

■ The allowance of 5 s before the emergency lighting must be in operation (it can be 15 s for some areas at the local authority's discretion) is based upon practical tests that indicated that after 15 s of darkness people began to move irrationally. Fortunately many of the specialist emergency luminaires are capable of virtually instantaneous operation and the 5 s requirement is redundant in such cases.

Planning the layouts

Positioning of luminaires

The correct positioning of emergency luminaires is essential in order to provide a system that not only complies with the various legislative requirements but provides a safe and effective way of evacuating a building in the event of a mains failure. Therefore, apart from achieving the minimum light levels, various obstructions, hazards and routes must be covered.

Luminaires and signs should be positioned as follows:

■ To show exit routes and final exits from premises clearly. Signs should be illuminated either internally or from an adjacent emergency luminaire.
■ To ensure exterior areas of final exits are lit to at least the same level as the area immediately inside the exit to enable people to move away from the exit to areas of safety.
■ Near each intersection of corridors (less than 2 m horizontally).
■ Near each change of direction (less than 2 m horizontally).
■ Near each staircase so that each flight of stairs receives direct light (less than 2 m horizontally).
■ Near any other change of floor level which may constitute a hazard (less than 2 m horizontally).
■ To illuminate fire alarm call points and fire fighting equipment at all material times.
■ To ensure normal pedestrian escape routes from covered car parks are illuminated to the same standard as escape routes within buildings.
■ In plant, switch- and control rooms.
■ Within passenger lift cars (only self-contained emergency luminaires are suitable for this application).
■ In toilets exceeding 8 m² or without daylight, and in all toilets for the disabled.

Self-assessment task 7.3

Would you normally provide emergency lighting in:

(a) a small private office
(b) a toilet 3 m × 2 m
(c) a reception area?

Table 7.17   Spacing for a surface-mounted luminaire for 1 lx minimum along the centre line

Ceiling mounting height (m)	Transverse to wall (m)	Transverse spacing (m)	Axial spacing (m)	Axial to wall (m)
2.5	1.8	5.6	4.7	1.5
3.0	1.5	5.5	4.6	1.2

Table 7.18   Spacing for a surface-mounted luminaire for 0.5 lx minimum in the core area

Ceiling mounting height (m)	Transverse to wall (m)	Transverse spacing (m)	Axial spacing (m)	Axial to wall (m)
2.5	2.1	5.6	4.6	1.7
3.0	2.0	5.8	4.8	1.7
4.0	1.7	5.8	4.9	1.5

In order to meet the illuminance levels manufacturers should provide a table of recommended spacings. A typical example is shown in Table 7.17.

Open areas

Emergency lighting is required for areas larger than 60 m^2 or open areas with an undefined escape route passing through. A minimum of 0.5 lx is required for the core area, i.e. it excludes the perimeter width of 0.5 m.

An open area can be illuminated either with separate emergency luminaires or with emergency packs fitted into standard luminaires. In either case the supplier should provide spacing data to meet the 0.5 lx minimum value. Table 7.18 is an example of the data needed.

Maintained operation

This system entails the continuous use of the emergency lighting installation which may be so designed that it appears, under normal conditions, to be part of the conventional lighting installation. It has the advantage that the performance of the emergency lighting is continuously monitored and any lamp or other failure should be instantly noted and dealt with before an emergency situation can arise. As it is usual to supply the emergency system from the normal supply under non-emergency conditions, the fact that the emergency lamps are working satisfactorily does not necessarily mean that the emergency power supply is operational and this must be independently checked.

The alternative power supply can be provided either by a prime mover generator set or by batteries. If a prime mover generator is used it must be capable of being started and run up to operating speed within 5 s of the failure of the normal supply. To meet such a requirement it may have an automatic starting device which will detect a failure of the normal supply. It requires a high degree of reliability to guarantee that it can be started and run up within the time specified. As a safeguard, it is possible that a bridging battery can be used to cover the period between failure of the normal supply and the start-up of the generator, which could then be done manually if desired. The British Standard recommends that a battery capable of at least 1 h duration should be used.

Where batteries alone are used, two methods of connection are possible. The first is the floating maintained system where the lighting load, battery and battery charger are connected in parallel and fed from the normal supply. Should there be a supply failure the battery will continue to feed the emergency lighting installation without the need for any change-over device.

The second is the maintained change-over system. Here the emergency lighting and the battery and charger are separately connected to the normal supply, no load being connected to the battery. In the event of a supply failure an automatic change-over device connects the emergency lighting to the battery. When the normal supply is restored the system reverts to its previous status and the battery is recharged.

Non-maintained operation

This is a system where the lamps are not normally in use but are automatically brought into operation in the event of a normal supply failure. In this case the emergency lamps may either be contained in a separate luminaire or be incorporated in the normal luminaire which may also house the battery, charger and change-over device for the emergency system.

Duration

It is recommended that an absolute minimum duration of 1 h should be provided for even the smallest premises. In some types of premises such as large hotels – particularly those on main thoroughfares where decanting space may be severely limited – it may be desirable to reoccupy the premises immediately the emergency is past or even to delay evacuation if this should be permitted. For these and other reasons a duration of 3 h is recommended for all premises having more than 10 bedrooms and more than one floor above or below ground level. Smaller premises should have a duration of 2 h.

As charging periods of 14 h are common for many contained units, problems can arise if the emergency system is exhausted in the early evening and people cannot return until the batteries are recharged or dawn provides the required minimum of 1.0 lx. Such problems are most likely to be solved by discussions with local authority officials and the local fire officer.

Location of exit signs (Pr EN 1838)

Signs are required at all exits, emergency exits and escape routes such that the position of any exit or route is easily recognised and followed in an emergency.

Fig. 7.12 Exit sign

The signs should confirm to the Health & Safety Executive Regulation (L64). Figure 7.12 is an example of such a sign. The maximum viewing distance for an internally illuminated sign is 200 × the height of the sign (*H*), and for an externally illuminated sign is 100 × *H*. An externally illuminated sign is a flat panel within 2 m of a standard emergency light.

EXERCISES

7.1   A factory floor requires lighting to 500 lx. The floor area is 20 m × 40 m and the ceiling height is 4 m. The wall reflectance is 30 per cent and the roof is glazed.

Using the data in Table 7.3 and making reasonable assumptions on maintenance, design a suitable lighting scheme. The work involves the painting of chinaware.

7.2   Explain the CIE method of assessing discomfort glare.

What factors does it have in common with the CIBSE glare index?

7.3   Using the data in Table 7.18 calculate the number of emergency lights for the factory floor given in exercise 1.

CHAPTER 8

LIGHTING OF SPECIFIC BUILDING TYPES

TOPICS COVERED

FACTORIES
  Illumination levels
  Glare
  Colour
  Luminaires
  Mounting
  Layouts

OFFICES
  Lighting of visual display screen (VDS) areas
  Glare control

SHOPS AND STORES
  Colour rendering
  Luminaires
  Layouts

THE HOME
  Illumination
  Layout

The guidance given in the preceding chapters applies to any lighting situation. However, different types of building and activity may require lighting solutions relevant to that building. If this were not so, then the only elements of choice would be based on economics. But it is not practical or even desirable to use the same solutions for all situations.

The CIBSE 1994 Lighting Code devotes 45 pages to recommendations for various situations and the purpose of this chapter is to outline the differences in lighting a factory, office, shop or home. It also indicates some of the sources of recommendations and further information.

FACTORIES

The principal requirements are to provide lighting for visual efficiency, safety and economy. The latest practice is contained in the CIBSE Lighting Code and the CIBSE Lighting Guide No. 1 (1989) *The industrial environment.*

Illumination levels

The Code contains detailed recommendations for the main industrial tasks and processes. These are directly related to visual tasks and examples are given in Table 8.1.

Glare

The limiting UGR range from 19 to 28 and glare control is, in general, less stringent than for offices. Reflected glare can cause problems when using high-pressure lamps in open-type reflectors.

Table 8.1   Typical recommendations for industrial premises

Process	Maintained illuminance (lx)	Limiting UGR
Boiler house	100	25
Loading bay	150	—
Unpacking store	200	25
Cable manufacture	300	25
Medium assembly	500	22
Paint spraying	750	22
Colour matching	1000	19

Source: Chartered Institution of Building Services Engineers, CIBSE 1994 Interior Lighting Code

Colour

A high standard of colour rendering is not normally needed. CR group 3 or 4 will be sufficient. However, where critical colour judgement is essential group 1A or 1B lamps must be used.

Luminaires

Open reflectors, with or without louvres, are most commonly used. Efficiency takes priority over aesthetics. If the environment is 'hostile' then special types are needed, e.g. flameproof, corrosion proof. These have already been outlined in Chapter 5. Further information is contained in the CIBSE Guide *Lighting in hostile and hazardous environments.*

Mounting

Ceiling heights can vary from 3 m to over 10 m. As the height increases it is possible to use the higher-wattage discharge lamps such as SON and MBI. When mounted at 10 m or more, the luminaires may be referred to as *high-bay.*

Layouts

In most cases a 'general' layout is used, as the factory may well be rented, and the lighting is installed before the plant layout is known.

Where high illumination levels are needed, such as for fine assembly work, it can be more effective to have a general level of 300–500 lx with local lighting boosting the assembly area to 750 lx. Where flexibility is needed then the lighting can be mounted on an electrical trunking system.

OFFICES

As with factories, visual efficiency is an important factor, but there is greater attention to glare, both discomfort and reflected disability. In general ceiling heights are 3 m or less and the lighting is often recessed within a suspended ceiling system.

The appearance of the scheme is more significant and the major proportion of office lighting uses fluorescent lamps. The main illumination and glare recommendations of CIBSE are given in Table 8.2.

Planning the lighting is similar to that for any general workspace. With the growing use of computers a different situation is created and this is the subject of EC display screen directive 90/270/ECC. The British approach is contained in the CIBSE Lighting Guide No. 3 (1996) *The visual environment for display screen use.*

Table 8.2   Typical recommendations for offices

Location	Maintained illuminance (lx)	Limiting UGR
General offices	500	19
Computer workstations	300–500	19
Drawing boards	750	16

Source: Chartered Institution of Building Services
Engineers, CIBSE 1994 Interior Lighting Code

Lighting of visual display screen (VDS) areas

There are two main visual concerns. The first is reduction in contrast in a display
screen due to reflections of the surrounds including the luminaires – a form of
veiling disability glare. The second is the need for a good lighting level to read data
being put into the VDS. Solutions to the two problems can be conflicting. One
system that appears to give satisfaction is to provide a general lighting level of
around 300 lx with strict control on glare. Where data have to be typed in, a
separate, well-screened, local light can provide additional illumination so that there
is at least 500 lx on the task, but not the screen.

Glare control

As already stated, the glare in this case is not discomfort but disability and the
restrictions are on luminances likely to be reflected in the screen. There are three
categories which refer to the situation and the control on luminaire design. The
control relates to luminances of less than 200 cd/m² above specified angles. This is
a low level and may be difficult to achieve.

The following are the definitions of the categories, quoted directly from CIBSE
LG3 (1996).

Category 1 luminaires

For Category 1 luminaires the calculated luminance is limited to 200 cd/m² or less
at and above 55° to the downward vertical. Such luminaires would be specified for
screens containing safety-critical or similar information where errors have serious
consequences. They may also be required in areas where there is a high density of
screens and display screen usage is of an intensive nature or sustained over long
periods.

Category 2 luminaires

For Category 2 luminaires the calculated luminance is limited to 200 cd/m² or less
at and above 65° to the downward vertical. Category 2 luminaires would, in general,

be slightly more efficient than Category 1 while maintaining good luminance control. This category of luminaire should be used in an interior for which fairly widespread use of display screens is intended. This could include areas where there is one terminal per desk for general usage or a few terminals used continually.

Category 3 luminaires

This is the greatest relaxation of the luminance limiting angle that can be recommended for areas where display screens may be used. The calculated luminance is limited to 200 cd/m^2 or less at and above 75° to the downward vertical.

By the nature of the relaxed control compared will the other categories they are, in general, more efficient and are capable of wider spacing. They should only be used for areas where the task makes only casual usage of the VDS and where the density of screens within an area is relatively low.

Some care may be needed in the positioning of individual screens within large open-plan areas.

Relaxation of luminance values

The figure of 200 cd/m^2 is very exacting and is based on viewing the worst kind of screen shape and surface. This includes negative polarity (light figures on a dark background). Reflections can be reduced by using positive polarity (dark figures on a light background) and anti-reflective surface treatment. If one of these is used the limit can be raised to 500 cd/m^2, and if both, 1000 cd/m^2. However, the lighting designer may well not know these facts and will have to use luminaires which conform to the 200 cd/m^2 value.

Exceptions

Category 2 and 3 luminaires are acceptable in higher-category applications where the space planning is either small cellular offices, or open plan with screen dividers, where by simple geometrical checking it can be shown that the luminaires will not be seen at angles below their limiting angle from the VDS.

The onus is on the manufacturers to state that their luminaires meet the requirements of a specific category. The designing of the layout follows the principles of any general layout and normally involves the use of a louvred design, of greater complexity for the lower categories.

One solution is to use uplighters. These have already been discussed in Chapter 7. The direct reflections are eliminated but the ceiling luminance may cause concern. LG3 recommends an average ceiling luminance of less than 500 cd/m^2. This is not difficult to meet, but the Guide also recommends a maximum luminance of less than 1500 cd/m^2. If the ceiling cavity has any glossy surfaces this may be impossible to achieve. Also with relatively low ceilings (less than 3 m), the luminance directly above the uplighter may well exceed the limit.

The health and safety regulations are outlined in Appendix B.

Table 8.3   Typical recommendations for retailing

Type	Maintained illuminance (lx)	Limiting UGR
Department store	750	19
Hypermarket	1000	22
Small shop	500	19
DIY store	1000	22
Family restaurant	200	19
Covered arcades and malls	50–300	22

Source: Chartered Institution of Building Services Engineers, CIBSE 1994 Interior Lighting Code

SHOPS AND STORES

Lighting for retailing covers a wide range of requirements and solutions, from the industrial approach for the large hypermarkets to a theatre lighting technique for the dress shop. In all cases a good level of illuminance and colour rendering on the merchandise is essential. Glare control is more arguable from the viewpoint of customers. They expect a reasonably bright interior with high lighting of the sales areas. Often a degree of flexibility is needed to allow for the display and sales layout to be altered. The CIBSE recommendations in Table 8.3 relate to the general lighting and stress the importance of vertical plane illuminance. This may be difficult to achieve from the general lighting and will need additional local display lighting.

There are no current CIBSE guides on the subject. Number 6, *The outdoor environment*, gives help in relation to public areas in shopping malls.

Colour rendering

Where a good standard is needed, e.g. clothes, food, furnishings, group 1B is required. This includes the triphosphor range of fluorescent lamps, and with their high efficacy there is a good case to use them throughout the retail area. Where ceilings are high and industrial luminaires for high-pressure lamps are used, as in DIY stores, then group 3 or 2 should be installed. This would include the SON deluxe and white SON (low wattages), MBF and metal halide lamps. The choice would depend on economic factors, the standard of colour rendering required, and the preferred colour appearance.

For premises over 100 m^2 the choice of lamps is influenced by the Building Regulations – see Appendix B.

Luminaires

Literally any type may be used depending on the style of shop and what is being sold. Many of the smaller shops use low-voltage track systems. Maybe 'misuse' is a truer description because the systems do need to be carefully sited and directed to illuminate the merchandise and not shine into people's eyes.

The larger stores often have a house style and you can recognise the store by its lighting. Most stores use suspended ceiling systems and the luminaires are recessed. However, the interior can look gloomy, despite satisfactory illuminance levels, unless there is a reasonable amount of light on the walls and ceilings.

Layouts

In all but the smallest shops a basic general layout is necessary. There may be many obstructions such as counter displays and cabinets and the illumination of vertical surfaces can be poor. Additional lighting is essential, either as part of the display feature or as separate spotlighting. It will be very difficult to calculate the illuminances that will be achieved in this way. Beam diagrams can help (see Chapter 2), and the important factor is to provide the electrical outlets for this additional lighting.

Finally, the lighting creates part of the shop's image. If the shop is a cheap and cheerful bazaar, then batten fluorescent luminaires could look right. If it is a department store, then the lighting should be unobtrusive but of a good standard. A high-class dress shop may well use crystal glass chandeliers with well-concealed additional lighting to reveal the expensive merchandise in a dramatic fashion. Whatever scheme is used, the customer should be aware of the merchandise and not the lighting.

THE HOME

The CIBSE 1997 Lighting Guide No. 9 contains recommendations for residential buildings, but these are for institutional rather than private homes. This is indicative of the difficulty in making meaningful suggestions on home lighting. The problem usually lies with the builder or developer, who, unless pressed, will provide the minimum in lighting points and socket outlets. The home owner will be limited to providing shades for single-centre light points and a few floor and table standard lamps.

Illumination

It is not necessary or even desirable to work to the same levels as experienced at work. However, a reasonable standard should be achieved and these are shown in Table 8.4. No UGR values are given because, although excessive glare should be avoided, the calculations cannot be applied to small rooms.

As with glare, it is virtually impossible to carry out illuminance calculations. As a very rough guide, filament lamps will provide approximately 100 lx per 30 W per m^2. Using fluorescent lamps this will fall to 8 W/m^2.

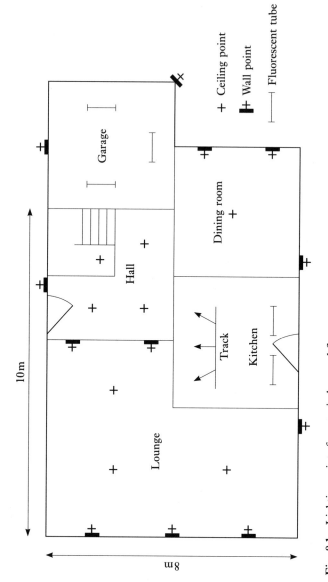

Fig. 8.1 Lighting points for a typical ground floor

Garage

Hall

Dining room

Lounge

Track

Kitchen

10m

8m

+ Ceiling point

+ Wall point

| Fluorescent tube

Table 8.4   Typical recommendations for residential rooms

Activity	Maintained illuminance (lx)
Reading	150
Cooking	150–300
Bedroom	100
Bathroom	150
TV viewing	50
Dining	150
Halls, corridors	100

Source: Chartered Institution of Building Services Engineers, CIBSE 1994 Interior Lighting Code

Layout

It is important to have sufficient lighting points. Figure 8.1 shows a possible lighting layout for the ground floor of a house.

Various points to take into account are:

- Lounge – central lights are normally reflected in TV screens.
- Kitchen – central lights cause shadows on most work surfaces.
- Bedroom – each bedhead should have its own well-shielded light.
- Outdoor – both front and back doors need external lights for safety and security.

Two books on home lighting may be of help: *Home Lighting*, by J. B. Harris, 1984; *Lighting*, by E. Effron, 1986.

This chapter has dealt with four general types of buildings. There are many more specialised types, such as museums and hospitals, and there are CIBSE lighting guides for many of these. The current guides are listed in the bibliography.

EXERCISES

8.1   A factory production floor is 30 m × 20 m with an 8 m ceiling height. The work is medium assembly of electronic components. Colour rendering must be group 3 or better.

Compare the merits of a high-pressure lamp scheme mounted at 8 m with a tubular fluorescent scheme mounted at 4 m. Assume both schemes have a utilisation factor of 0.7 and indicate typical layouts.

8.2   Discuss the particular problems in lighting:

(a) a chemist's shop and dispensary
(b) a railway station platform
(c) a jeweller's shop

Items to be considered include illuminance levels, glare, colour, types of lamp, maintenance and protection.

CHAPTER 9

ENERGY MANAGEMENT AND LIGHTING

TOPICS COVERED

DESIGN ELEMENTS
Installed lighting loads

TYPES OF LAYOUT
Planned maintenance
Lamp efficacies
Lighting control systems
Heat recovery from luminaires
Energy saving developments

To obtain the best performance from a lighting installation in terms of energy consumption, and hence cost, it must be efficiently designed, used and maintained. This chapter introduces a number of topics which must be considered, and while each can produce a saving, all the elements should be considered if the saving is to be worthwhile.

DESIGN ELEMENTS

The various elements can be summarised as follows:

- The highest utilisation factor (UF) compatible with the lighting design requirements.
- A layout which provides higher illuminance values only where they are required.
- The highest maintenance factor (MF) compatible with a realistic maintenance programme.
- The use of appropriate lamps with the maximum luminous efficacy.
- Switching control system that enables full use to be made of daylight.
- If the building is suitable, the heat generated by the lighting system to be integrated into other building service requirements.
- The lighting to be used only when required.
- Advantage to be taken of improvements in lamp efficacies as they occur.
- The installation should last only as long as it remains efficient, i.e. until it can be proved cost effective to replace it with a more up-to-date installation.

Installed lighting loads

The load can be expressed in watts per square metre. There are too many variables for a precise guide on good design to be given, but Table 9.1 gives an indication of values that could be expected for different room indices. The values of watts include the total circuit watts, i.e. the power consumed by the lamp and its control circuit.

Table 9.1   Guide of target wattage for general lighting

Type of interior	Maintained illuminance (lx)	Room index	Utilisation factor (guide)	Target load for circuit efficacy 60 lm/W (W/m²)
Heavy industrial	500	5	0.7	15
		2	0.63	16.5
Light industrial	500	5	0.6	17.5
		2	0.53	20.0
Commercial	500	5	0.5	21.0
		2	0.42	25.0

Example

Referring to Chapter 7, does the lighting scheme designed for the drawing office (see Fig. 7.2) meet the target?

Solution

The loading is 12 65 W white fluorescent tubes. The control gear will add about 15 per cent to the loading. Therefore,

$$\text{Total} = 12 \times 65 + 12(0.15 \times 65)$$
$$= 897 \text{ W}$$

The area is 6 m square. Therefore,

$$\text{Loading} = \frac{897}{6 \times 6}$$
$$= 25 \text{ W/m}^2$$

This is to produce 750 lx with a circuit efficacy of 5200 lm per 74 W lamp circuit or 70.2 lm/W. The room index is 1.4. Referring to Table 9.1, the commercial target load for a room index of 2 is 25 W/m^2. Extrapolating to a room index of 1.4 this becomes 25.5 W/m^2. Therefore,

$$\text{Target load} = 25.5 \times \frac{750}{500} \times \frac{60}{73.2}$$
$$= 31.3 \text{ W/m}^2$$

This indicates that the scheme proposed is well within the target.

Self-assessment task 9.1

What would the load have been if the same space was lit with 36 50 W halogen spots on a 1.2 m grid to provide 400 lx? Would it meet the target for a commercial area?

TYPES OF LAYOUT

Various layouts have already been considered in Chapter 7. The local layout may provide the best opportunity for energy saving as it only provides higher illuminances in the task area with a lower level of general illuminance elsewhere.

An interesting development is the use of 'uplighters' to provide the general illuminance. Such a scheme is shown in Fig. 9.1 and this permits the use of efficient high-pressure lamps such as SON deluxe and MBIF in general office situations. Table 9.2 shows comparative loads for this type of installation and an equivalent general lighting scheme. There are critics of uplighter schemes. The ceilings are too bright for some; the MF tends to be low unless the maintenance is of a high order; and a single lamp failure causes a fair degree of disruption. However, it is a system which is the subject of considerable interest and research.

Table 9.2 Comparison of general lighting scheme with uplighter for a general office space

Scheme	Loading (W/m²)	Illuminance (lx)
General	17	500
Uplighter/desk light	approx. 12	250/500

General scheme: Conventional ceiling-mounted fluorescent luminaires to provide 500 lx
Uplighter scheme: 250 W SON deluxe lamps in uplighters to provide 250 lx general
lighting; 18 W fluorescent tubes for desk lighting to 500 lx

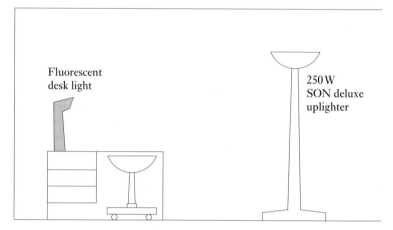

Fluorescent
desk light

250 W
SON deluxe
uplighter

Fig. 9.1 Uplighter system for an office

Planned maintenance

The MF has as much effect as the UF on calculating installed load, and if the value is to be any more than a pious hope, it should be linked to a planned maintenance schedule. Unfortunately the engineer who selects the MF is seldom responsible for establishing a maintenance schedule.

Figure 9.2 illustrates the pattern of light loss due to lamp depreciation and lamp and luminaire soiling. This will vary, depending on the characteristics of the environment, luminaire and lamp. The figure illustrates that a systematic cleaning and lamp replacement programme will make a significant improvement to the overall performance of the lighting.

Example

Using the data in Fig. 9.2, what will be the MF after 9000 h of use if the lamps are replaced after 6000 h and the cleaning interval reduced to 1500 h of use?

Solution

This is illustrated in Fig. 9.3 and is 70 per cent, a considerable improvement on 51 per cent in Fig. 9.2.

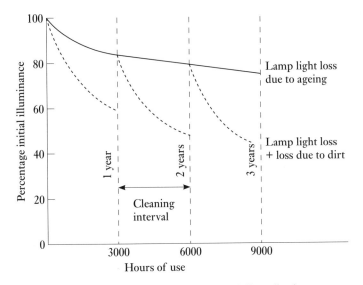

Fig. 9.2   Typical light loss curves due to lamp ageing and dirt collection

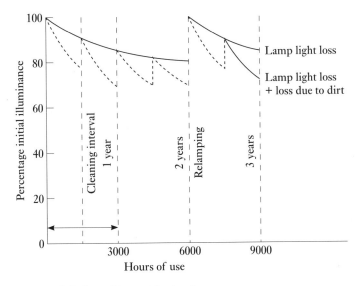

Fig. 9.3   Effect on light loss of 6 monthly cleaning

The systematic cleaning and lamp replacement is known as planned maintenance; it is not always a practical exercise, but where it is used the lighting is more efficient, visually more satisfactory, and can be carried out with the minimum of disruption.

Table 9.3 Comparative lamp efficacies and costs

Lamp type	Typical luminous efficacy (lm/W)[a]	Approx. cost[a] (£/100 W)	Control gear cost (£)
GLS	12	0.5	—
TH	22	1.6	—
SOX	180	20	40
SON	100	18	25
MBF	55	6.5	12
MCF	90	2.8	8
MBI	75	14	25

[a] These values are based, where possible, on lamps of about 100 W rating.

Self-assessment task 9.2

Looking at Fig. 9.2, find the total light loss after 3 years if the cleaning interval is every two years.

Lamp efficacies

Lamp choice is related to: colour appearance; colour rendering; lumen output; physical size; cost; and life. Any of these factors may be the overriding consideration limiting the choice to lamps which would otherwise be considered inefficient. The tungsten-filament lamp is the only choice for use in a crystal chandelier. Likewise, however high the efficacy of a low-pressure sodium lamp, it will not be accepted for office lighting. So lamp selection can seldom be only on the grounds of high luminous efficacy. Table 9.3 lists the range of lamps already discussed in Chapter 4 in terms of their luminous efficacy and average cost per 100 W.

Lighting control systems

The aspects discussed in this chapter, so far, refer to savings that can be achieved by using energy-efficient equipment. However, even the most efficient lighting is wasteful if it is in use when no one needs it. Intelligent control could save up to 70 per cent of the lighting load, but 25 per cent is a more realistic figure.

Occupancy control

Occupancy control detectors can be used to switch lights on as people enter a room and off again after they have left. This avoids lights being left on unnecessarily. They can be used to operate task lighting or lighting in rooms which are used infrequently, such as store rooms where people are likely to have their hands full on entering.

The circuit will need to include a time delay to allow people to leave the space safely and to avoid lights being constantly switched on and off.

Presence detectors can be ceiling or wall mounted, but the sensor must be able to detect an occupant at all times. Sensors must be sufficiently sensitive to operate when required, but not too sensitive that they respond to extraneous signals.

An option is to combine a presence-operated switch with a manual switch. The occupant switches the lights on manually when required, and the presence detector switches them off. This is sometimes referred to as 'absence sensing'.

Presence sensors can be used for switching lights on and off when security and cleaning staff move around a building outside normal hours, and for security purposes to signal the presence of an intruder.

Electric/daylight control

The relative roles of daylight and electric light have already been discussed in Chapter 6.

In areas where there is adequate daylight for part of the time, daylight illuminance sensors (photocells) can be used to ensure that electric lights are not left on unnecessarily. People will often switch electric lights on first thing in the morning when it is still dark, but they are less likely to switch them off later when daylight becomes sufficient, particularly in shared spaces and circulation areas. Illuminance sensors can switch or regulate luminaire light output, but dimming will require the appropriate control gear.

Self-assessment task 9.3

For what range of average daylight factor values is it worth considering some form of daylight/electric light control?

Planning a system

The main problem is cost. Will the cost of the control justify the energy saving achieved? What about 'ease of use'? Is the system user-friendly or will it be too complex to be properly operated and maintained? If a dimming control is to be used, are the lamps and their control circuits suitable?

There are a number of companies who can advise and supply control systems and, as a starting point, the Department of the Environment has produced a 'Good Practice Guide 160 – Electric lighting controls'.

Processor control

Computer- or microprocessor-based control systems are becoming increasingly popular, reliable and less expensive. These rely upon dedicated computers or processors to control some or all of the building services. An advantage of such an approach is that complex decisions can be taken from moment to moment, based upon the precise state of the building's operation. The system is controlled by software, which means that the control programs can be refined and tailored to suit the building and can be amended to suit changed circumstances.

Table 9.4   Energy distribution of light sources

Light source	Convected/conducted (%)	Radiant (%)	Source temp. (°C)
Fluorescent tube	51	49	40
MBF	30	70	300–800
HPS/SON	26	74	300–800
Tungsten	15	85	2500

Table 9.5   Lamp and circuit wattages

Lamp type	Nominal lamp wattage	Approx. total circuit wattage
Fluorescent (high frequency)	58	67
Mercury MBF	125	145
	400	435
Metal halide MBI	250	291
	400	445
Sodium SON	250	310
	400	450

With any control system considerable care must be taken to ensure that acceptable lighting conditions are always provided for the occupants. Safety must always be of paramount importance.

Heat recovery from luminaires

Lamps are comparatively inefficient producers of light, but 100 per cent effective producers of heat. In buildings lit to modern standards, the heat generated by the lighting can make a significant contribution to the heating of the building. It can also be an undesirable heat gain for a refrigeration plant.

Table 9.4 compares the generation of heat by various light sources. In the case of all discharge lamps, a further 7–20 per cent of conducted and convected heat must be added for the control gear. When the lamp wattage is quoted, it does not usually include the control gear. Table 9.5 shows some typical lamp and total circuit wattages.

Irradiance from lighting

As already shown, there is a significant amount of radiant power which is a mixture of ultraviolet, light and infrared. There is little that can be done to control this heat and people are sensitive to radiant power, in particular on their heads and the backs of their hands.

Table 9.6 shows typical irradiance levels for different types of installation. The values vary quite considerably, depending on the distance from the luminaires. A person with a bald head will be uncomfortably aware of recessed filament lighting

Table 9.6   Typical irradiance values for different types of fluorescent lighting using white tubes

Mounting	Luminaire	Illuminance (lx)	Irradiance (W/m²)
Surface	Batten	500	3.6
	Opal diffuser	500	2.6
	Prismatic controller	500	4.5
Recessed	Opal diffuser	500	4.1
	Prismatic controller	500	5.0

in a low ceiling even though the average working plane illuminance may be only 100 lx. The irradiance cannot be eliminated and may even in some situations be desirable, but lamps radiating directly onto people at comparatively short distances should be avoided. An example of a situation needing careful design is the use of track systems in shops where spotlamps may be haphazardly located causing heat discomfort to the shop employees and the customers.

Effect of luminaire on heat distribution

Lamps radiate a large proportion of their energy, but when they are placed in luminaires a part of that energy is absorbed and converted into conducted and convected energy. The amount depends on the construction and optical system. Table 9.7 shows a range of typical luminaires and the approximate distribution of energy they emit.

Air-handling luminaires

Luminaires are available which are specifically designed to enable the room air to flow over the lamps, control gear and metal surfaces, extracting up to 75 per cent of the heat generated by the luminaire.

Figure 9.4 shows typical designs, but there are many variations depending on the method of integrating the air-handling components with the general mechanical services. The advantages of using such a system include:

- A more economic use of the energy within a building. For instance, refrigeration load is not required to counteract heat generated by the lamps.
- Fluorescent lamps run at a more efficient temperature. The effect of temperature on light output has already been discussed in Chapter 4 and improvements of up to 10 per cent can normally be expected.
- The visual environment is improved by the integration of ceiling services. Often where lighting and air handling are designed as separate services, their layouts conflict giving an untidy effect on the ceiling.

This form of heat recovery is usually feasible where the gross heat gain from the lighting, occupants and equipment is comparatively large and the interior is well insulated thermally. It may suit a deep office, but could be quite unsuitable for a shallow building with a high percentage of the exterior wall glazed.

Table 9.7 Energy distribution for different types of luminaires

Type of luminaire		Energy distribution (%)	
Mounting	Detail	Up	Down
Recessed	Open	38	62
	Louvre	45	55
	Prismatic or opal diffuser	53	47
Surface	Open	12	88
	Enclosed prismatic or opal	22	78
	Enclosed prismatic on metal spine	6	94

Method of installation

Air can be extracted from the room through the luminaire. This is shown in Fig. 9.5 and the system would be fully ducted. The ductwork for the return air can be reduced by exhausting through the luminaire into the plenum, as shown in Fig. 9.6. For this to work, the whole of the ceiling system must be reasonably airtight. Leakage of air through the ceiling can unbalance the system, cause ceiling tile pattern staining and reduce the heat transfer efficiency of the luminaire.

Performance of luminaires

Apart from the usual photometric data, a manufacturer of air-handling luminaires must supply sufficient data for the heating and ventilation calculations.

To obtain these data the luminaire must be tested in a calorimeter and the following is normally required:

- total heat flow to plenum
- rate of heat input (electrical load)
- exhaust air temperature
- room air temperature
- relative light output
- air flow rate

Test data are normally presented in graphical form and a typical set of results is shown in Fig. 9.7.

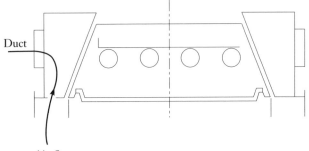

Duct

Air flow

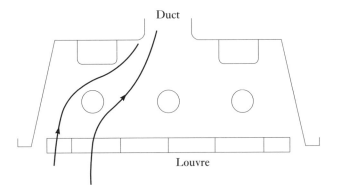

Duct

Louvre

Fig. 9.4 Two forms of air-handling luminaires

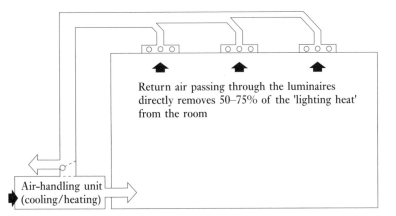

Return air passing through the luminaires directly removes 50–75% of the 'lighting heat' from the room

Air-handling unit (cooling/heating)

Fig. 9.5 Typical ducted installation of air-handling luminaires (courtesy The Electricity Association)

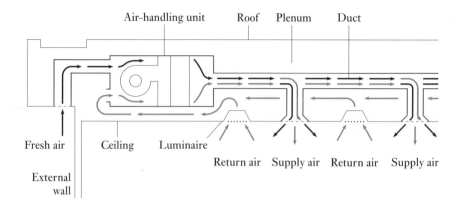

Fig. 9.6   Use of the plenum as an exhaust chamber (courtesy The Electricity Association)

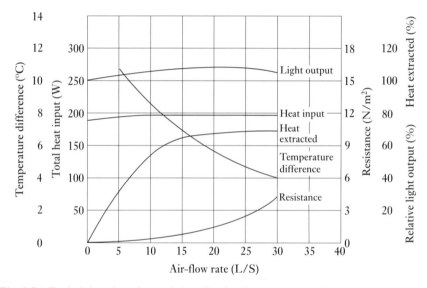

Fig. 9.7   Typical data sheet for an air-handling luminaire (courtesy Whitecroft Lighting Ltd)

Self-assessment task 9.4

Give three reasons for considering the use of air-handling luminaires.

Energy-saving developments

The lighting industry has produced many ideas to make effective savings of energy. Some have been little more than sales gimmicks, others have been of value. The aim is normally to achieve the same or better illuminance from an existing installation by using different lamps. The following are some of the more successful ideas.

Blended lamps (MBT or ML)

If GLS lamps are in use, improved efficiency normally requires replacement with discharge lamps. These need control gear, which implies new luminaires. Blended lamps are mercury discharge lamps incorporating a series resistance in the form of a tungsten filament. This results in a lamp of around 20 lm/W with a good standard of colour rendering. It is a compromise in that it is not as efficient as a standard MBF lamp, but it is a relatively inexpensive method of upgrading a GLS installation.

Plug-in lamps

High-pressure mercury lamps have been available since the mid-1930s. They have proved an economic and reliable light source but there are many installations using 250 W or 400 W mercury lamps which could be significantly upgraded in efficiency and light output. Table 9.8 compares the performance of the mercury lamp and its equivalent plug-in high-pressure sodium lamp. Savings of 15 per cent in power consumption and up to 50 per cent higher light output can be achieved by simply replacing the mercury lamp with the equivalent plug-in sodium lamp with no change to the electrical control circuit.

Fluorescent lamps

Developments in the lamp and its control circuit can achieve significant savings in power consumption. The savings made using high-frequency ballasts have already been discussed in Chapter 4.

Many luminaires are still using the T12 – 35 mm diameter lamps with argon gas filling. On switch-start circuits these can be replaced by T8 – 26 mm diameter lamps with krypton gas filling achieving an immediate energy saving of 8 per cent.

At the design stage, if triphosphor lamps are used in preference to halo-phosphate lamps then, as they produce 10 per cent more light and have a slower rate of lumen depreciation, they will save 10–15 per cent of the power needed to achieve a specific maintained illuminance.

Adding all these factors together, modern development allows an old fluorescent installation to be replaced with power savings of up to 50 per cent.

Table 9.8   Comparison of MBF and equivalent 'plug-in' SON

MBF mercury lamp (W)	Initial lumens	Plug-in SON-E lamp (W)	Initial lumens
250	13 000	220	22 500
400	22 000	310	24 500

CONCLUSION

As long as energy costs rise, energy-saving schemes will develop. It would be pleasing to think that energy-saving schemes are introduced to save energy. The reality is that they are only introduced to save cost, and if this is not achieved, the ideas are abandoned. Costing a scheme is a complex procëss involving energy tariff, interest rates and taxation methods. The best form of economy is to design an efficient and effective scheme and to maintain it properly.

EXERCISE

9.1   An office has an average daylight factor of 4 per cent from windows in one wall. There is a true ceiling of 3.2 m height below which is a suspended ceiling 2.8 m above floor level. Reasonable colour rendering is required.
(a) List the range of suitable lamp types.
(b) Suggest a form of control for economic use of available daylight.
(c) Calculate a target load for providing 300 lx in an office of 10 m $\times$ 20 m.
(d) Calculate the amount of heat (in watts) that can be usefully recovered. To do this you need to assume reasonable UF and MF values and select a lamp type.

CHAPTER 10

ROADWAY LIGHTING

TOPICS COVERED

SILHOUETTE VISION

GLARE

DESIGN RECOMMENDATIONS
 Part 1: Guide to the general principles
 Part 2: Lighting for traffic route
 Part 3: Lighting for subsidiary roads and associated pedestrian areas
 Outdoor light pollution

There are few people who do not appreciate a well-lit street, be they motorist, passenger, cyclist or pedestrian. They may not like it aesthetically, but that is another story. What do they appreciate?

■ The degree of safety in an otherwise extremely hazardous situation.
■ The ability to see the outdoor scene in total, not as a restricted view in the beam of a car headlight.

These could be summarised as 'the ability to see and be seen'.

How does one define 'a well-lit street'? It would have to appear evenly illuminated, with no obvious dark spaces and a minimum of glare.

How is this achieved? Not in the same way as a well-lit interior. There is no convenient 'ceiling' to fit the lighting to, no walls to reflect light and it would be impractical to maintain even illumination along the roadway. The design techniques developed for interior lighting are of little use in this situation.

SILHOUETTE VISION

When a road is referred to as being 'evenly lit', this means that, when viewed from a car, the road surface appears to be 'evenly bright'.

If a lamp is hung some 10 m above a road surface, a patch of light is reflected from the road. The shape of this patch depends on the road surface. On surfaces of very fine texture which take a noticeable polish, such as asphalt, the patch is long, extending even to the feet of the observer. On the more usual rough roads, the patch extends across the road rather than down it. Hardly any bright area will be seen on the far side of the post supporting the lamp (Fig. 10.1). The patches are not so well defined as shown in the sketch, but can nevertheless be observed quite distinctly.

If a succession of lanterns along the length of the road is so arranged that the bright patches merge to cover the road area, objects on the road will be seen as dark silhouettes against the bright surface (Fig. 10.2). This is the principle upon which most street lighting is based, since it proves more economic to produce silhouettes than it would be to make objects light and the road surface dark.

By day the illuminance on the road surface is approximately constant along the length. By night the reflection is quite different. The maximum illuminance on the road is immediately below the hypothetical bare source and tails off up and down the road, dropping to one-tenth the maximum at about 12 m and to a negligible value at twice this distance. The luminance of the surface, however, to an observer some 70 m down the road, has its maximum at about 12 m from the foot of the column, falling off steadily closer to the observer.

The long bright patch is usually described as 'T-shaped', having a 'head' and a 'tail'. The head is formed by light emitted from the lantern over the range of angles from vertically downward up to about 60° (i.e. 30° below the horizontal). Light above the 60° angle forms the tail.

Self-assessment task 10.1

What effect is rain likely to have on the luminance patches?

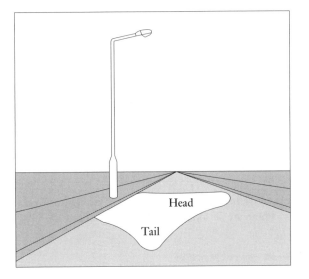

Fig. 10.1 Luminance patch from a single lamp

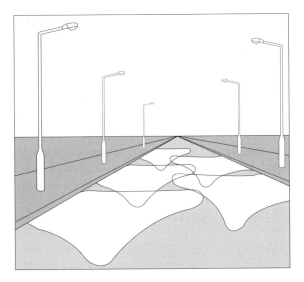

Fig. 10.2 Luminance patches merging to give even lighting

GLARE

One objection to lighting a main road with rows of powerful bare lamps would be the glare. Examination shows that much of this arises from light emitted between 70° and the horizontal, i.e. the light which produces the long tail.

If lighting highly polished roads or permanently wet ones, it may be worthwhile to accept this glare in return for the greater 'covering power' (and, therefore, longer spacing and cheapness) of the bare lamp. But since most roads are rough and more often dry than wet, this exchange is not generally worthwhile. Therefore, the lantern light distribution is so designed that only a little light emerges above 70° to the downward vertical.

The aims of good lighting are to provide a road surface of even luminance as seen by the road user, without an unacceptable level of glare from the lanterns.

DESIGN RECOMMENDATIONS

A very complex exercise has been reduced to fairly straightforward calculation and design processes. These are set out in the British Standard BS 5489: 1992 *Code of practice for road lighting*.

Part 1:   *Guide to the general principles*
Part 2:   *Lighting for traffic routes*
Part 3:   *Lighting for subsidiary roads and associated pedestrian areas*
Part 4:   *Lighting single-level road junctions, including roundabouts*
Part 5:   *Lighting for grade-separated junctions*
Part 6:   *Lighting for bridges and elevated roads*
Part 7:   *Lighting of tunnels and underpasses*
Part 8:   *Lighting for roads near aerodromes, railways, docks and navigable waterways*
Part 9:   *Lighting for town and city centres and areas of civic importance*
Part 10: *Lighting for motorways*

Part 1: *Guide to the general principles*

This is a relatively short section. Its purpose is to state the principles on which the other parts are based. It includes definitions, road classifications, daytime appearance, the visual scene and task at night, the intentions of road lighting, the hazards of columns, the law, maintenance.

In understanding the design process the following is of help.

Definitions

Arrangement   The pattern according to which lanterns are sited in plan, e.g. staggered, opposite, single side and twin central.

Average luminance (of the road surface)   The average luminance over a defined area of the road surface viewed from a specified observer position.

Design attitude (of a lantern)   The disposition of a lantern in space, usually indicated by a diagram or by reference to a datum axis such as the spigot entry.

Design spacing   The required spacing between lanterns, calculated as specified in the various parts of BS 5489, for a straight and level section of the particular type of road.

Effective width (W_E) A notional distance, related to the width on the carriageway, the lantern overhang and the arrangement, used to simplify the design tables.

Equivalent veiling luminance (L_v) Luminance that is, in effect, superimposed on the visual scene as a result of bright areas in the field of view.

Geometry (of a lighting system)   The interrelated linear dimensions and characteristics of the system, i.e. spacing, mounting height, effective width, overhang and arrangement.

Longitudinal uniformity ratio (of luminance) (U_l) The ratio of the minimum to the maximum luminance along a longitudinal line through the observed position on the carriageway.

Maintenance factor (MF)   The product of the lamp flux depreciation factor (LFDF) and the luminaire maintenance factor (LMF), i.e. $MF = LFDF \times LMF$.

Mounting height   The nominal vertical distance between the photometric centre of a lantern and the surface of the carriageway.

Overall uniformity ratio (of luminance) (U_o) The ratio of the minimum to the average luminance over a defined area of the road surface viewed from a specified observer position.

Overhang (A) The distance measured horizontally between the photometric centre of a lantern and the adjacent edge of a carriageway. The distance is taken to be positive if the lanterns are in front of the edge and negative if they are behind the edge.

Set back   The shortest distance from the forward face of a column to the edge of a carriageway.

Spacing   The distance measured parallel to the centre line of the carriageway, between successive lanterns in an installation.

Spacing index (S_l) The product of the calculated average road luminance (in cd/m^2) produced by an installation of lanterns in their clean state and the spacing (in m) divided by the lamp flux (in klm).

Surround ratio (S_R) The ratio of the average illuminance on a strip 5 m in width beside the carriageway to the average illuminance on the adjacent 5 m strip of carriageway.

Threshold increment (TI)   A notional measure of the effect of disability glare from street lanterns.

Width of carriageway (W_K)   The distance between kerb lines, measured at right angles to the length of the carriageway.

Road classification

The full classification is as described in Parts 2–10. The parts of main concern in this chapter are:

- Lighting for traffic routes (BS 5489: Part 2).
- Lighting for subsidiary roads (BS 5489: Part 3).
- Lighting for single-level road junctions including roundabouts (BS 5489: Part 4).

Part 2: *Lighting for traffic routes*

This is the 'meat' of BS 5489 as it contains the necessary instructions on the design of lighting for all-purpose traffic routes up to 15 m in width for single carriageways, and up to 2×11 m in width for dual carriageways. It does not deal with motorways. The following sections deal with the calculations but a copy of BS 5489 is essential for the complete design process.

Mounting height (H)

In general, increasing H enables wider spacing to columns. This may be of help on long open stretches of road, but not so in urban situations with many junctions and intersections.

   The choice of H is 8 m, 10 m or 12 m. A height of 10 m is the recommendation for the majority of roads, with 12 m for wide or heavily used roads such as major routes between two towns.

Control of glare

Disability is the criterion (rather than discomfort). It lends itself better to numerical treatment and it is considered that limiting disability glare also limits discomfort glare.

   Glare is a function of the source luminance and the luminance of the background. An object that is just visible (i.e. at the threshold of visibility) when there is no disability glare will, in the presence of disability glare, merge into the background. The percentage by which the background luminance has to be increased to render the object just visible again is known as the threshold increment (TI). This provides a notional measure of disability glare.

   The value of TI depends on the light distribution from the lantern between 70° and 90° in elevation in the vertical plane at which the lantern is observed. It also

Table 10.1   Recommended values of luminance and uniformity ratio (adapted from BS 5489 Part 2: 1992)

Category	Average luminance $\bar{L}$ (cd/m²)	Overall uniformity ratio U_o	Longitudinal uniformity ratio U_L	Examples
2/1	1.5	0.4	0.7	High-speed road, dual carriageway roads
2/2	1.0	0.4	0.5	Important rural and urban traffic routes
2/3	0.5	0.4	0.5	Radial roads, district distributor roads

depends on the road luminance, the layout of the lanterns, the mounting height and the observer position.

A value of TI not exceeding 15 per cent is recommended where it is necessary to minimise glare, e.g. on high-speed roads. It is of particular advantage in rural areas, where the absence of a reflecting background, such as buildings, may increase the effect of glare. In all other installations the value of the TI should not exceed 30 per cent.

Manufacturers normally classify their lanterns as LTI (Low Threshold Increment) to meet the 15 per cent requirements or MTI (Moderate Threshold Increment) to meet the 30 per cent value. The British Standard gives a method for working out the TI but suggests this will be needed only on long straight roads where a large number of lanterns are visible.

Surround ratio (S_R)

This has already been defined (p. 215) and it is a method of ensuring that there is adequate visibility for pedestrians and that vehicle drivers can see people and objects to the side of the carriageway. An S_R value of not less than 0.5 is suggested and this can be calculated from the isolux diagrams.

Luminance of road surfaces

The aim of good lighting is to achieve a road surface of even brightness. The evenness is expressed in terms of U_o (overall uniformity ratio – see definitions). This will depend on the type of road surface, lantern and the geometry of the layout. Another term U_L (longitudinal uniformity ratio) is also used. U_o is of such a value that no part of the road is too dark and U_L of a value so that the appearance is not too 'patchy'.

The average level of brightness is expressed in luminance (cd/m²). Table 10.1 shows the BS recommendations.

To achieve the values the layout of columns and lanterns (the geometry) is planned with the help of data and tables contained in Part 2.

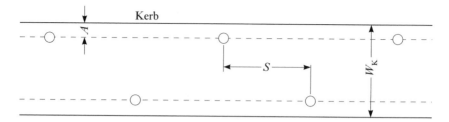

Fig. 10.3 Staggered arrangement of lanterns. *Note:* Effective width (W_E) is $W_K - 2A$ for lanterns over the carriageway and $W_K + 2A$ for lanterns behind the edge of the carriageway

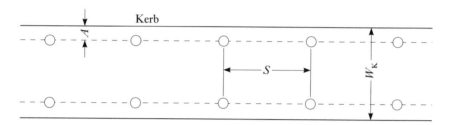

Fig. 10.4 Opposite arrangement of lanterns. *Note:* Effective width (W_E) is $W_K - 2A$ for lanterns over the carriageway and $W_K + 2A$ for lanterns behind the edge of the carriageway

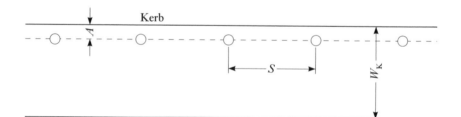

Fig. 10.5 Single side arrangement of lanterns. *Note:* Effective width (W_E) is $W_K - A$ for lanterns over the carriageway and $W_K + A$ for lanterns behind the edge of the carriageway

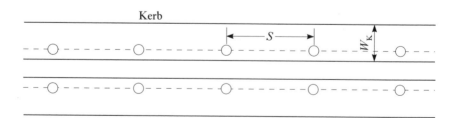

Fig. 10.6 Twin central arrangement of lanterns on dual carriageway. *Note:* Effective width (W_E) is W_K

Figures 10.3–10.6 show all the important dimensions and what is meant by a 'staggered', 'opposite', 'single-sided' and 'central' layout.

To help follow the design steps, Table 10.2 shows a typical set of a manufacturer's photometric data for a lantern suitable for main traffic route use. This is to be used to light a category 2/2 road whose carriage width is 10 m. At this stage junctions and bends will not be considered.

Road surface

Luminance values of the road surface depend on the light received and the reflectance properties of the road surface. The latter comprise both the total reflectance and the specular reflectance at various angles. The total reflectance is mainly determined by the colour of the stone chippings, and the specular properties by the type of surface finish. A range of types are defined in CIE Publication No. 66 1984: *Road surfaces and lighting*.

However, the range of surfaces is so wide that the British Standards Institution, in their wisdom, has defined a 'representative British road surface' as typical of the great majority of road surfaces used in the UK. The average luminance coefficient for this surface is 0.07 and the full reflection data are given in BS 5489. Data issued by manufacturers are based on this surface. With the help of a computer it would be possible to produce data for other surfaces. The CIE classification of this surface is C2.

Geometry of the layout

Assume a staggered layout with a 1 m overhang (A). Also assume that MTI will meet the glare control.

Next refer to Table 10.2 to check that S_R meets the requirements of 'not less than 0.5'. The value in the table will be for W_E of 8 m (Fig. 10.3) and this is 0.85, which is satisfactory.

Next the spacing index S_I is seen to be 3.38. This index is defined as

$$S_I = \frac{S\overline{L}}{\phi \times MF} \qquad [10.1]$$

where S is the spacing in metres, $\overline{L}$ the average road luminance (in cd/m^2), ϕ the initial design lumens (in klm) and MF the maintenance factor (see Tables 10.3 and 10.4).

This formula can be rearranged to give an actual spacing S:

$$S = \frac{S_I \times \phi \times MF}{\overline{L}}$$

The initial lumen output of a 150 W SON plus lamp is 16 klm (Appendix C). The MF is to be based on:

Table 10.2 Design table: standard presentation

Lantern
Staggered

Lamp: 150 W SON/T
Design attitude: Spigot entry elevated 5°
Mounting height: 10 m

Lantern classification MTI

S (m)	W_E 6 m S_I 3.78 S_R 0.83			W_E 7 m S_I 3.58 S_R 0.84			W_E 8 m S_I 3.38 S_R 0.85			W_E 9 m S_I 3.19 S_R 0.85			W_E 10 m S_I 3.02 S_R 0.86		
	U_o	U_L	V_F	U_o	U_L	V_F	U_o	U_L	V_F	U_o	U_L	V_F	U_o	U_L	V_F
20	0.68	0.81	29	0.68	0.80	29	0.67	0.81	27	0.66	0.83	27	0.66	0.83	26
22	0.66	0.81	27	0.65	0.80	26	0.65	0.80	26	0.65	0.82	25	0.65	0.82	25
24	0.65	0.82	25	0.64	0.81	24	0.65	0.81	24	0.65	0.81	24	0.66	0.84	23
26	0.62	0.73	24	0.62	0.73	23	0.62	0.72	23	0.63	0.73	23	0.63	0.74	22
28	0.63	0.75	22	0.62	0.74	22	0.63	0.71	22	0.64	0.70	22	0.63	0.70	22
30	0.63	0.73	21	0.63	0.73	21	0.63	0.71	21	0.61	0.71	20	0.59	0.71	20
32	0.63	0.67	21	0.63	0.65	21	0.61	0.63	20	0.58	0.62	19	0.57	0.62	19
34	0.63	0.67	19	0.60	0.63	19	0.58	0.61	19	0.56	0.61	19	0.54	0.58	18
36	0.61	0.67	19	0.56	0.63	18	0.54	0.58	18	0.50	0.56	18	0.49	0.55	18
38	0.59	0.64	18	0.58	0.61	18	0.54	0.58	18	0.49	0.56	17	0.47	0.53	17
40	0.55	0.58	17	0.53	0.56	17	0.51	0.54	17	0.50	0.52	16	0.48	0.50	16
42	0.54	0.57	17	0.52	0.54	17	0.51	0.52	16	0.50	0.50	16	0.47	0.48	16
44	0.56	0.56	16	0.53	0.50	16	0.51	0.51	16	0.49	0.49	16	0.46	0.48	15
46	0.55	0.56	16	0.52	0.52	16	0.50	0.50	15	0.48	0.49	15	0.46	0.47	15
48	0.53	0.55	16	0.50	0.51	15	0.47	0.49	15	0.46	0.47	15	0.45	0.45	14
50	0.52	0.50	15	0.49	0.46	15	0.46	0.44	14	0.45	0.42	14	0.44	0.40	14

S (m)	W_E 11 m S_I 2.84 S_R 0.88 U_o	U_L	V_F	W_E 12 m S_I 2.71 S_R 0.90 U_o	U_L	V_F	W_E 13 m S_I 2.62 S_R 0.92 U_o	U_L	V_F	W_E 14 m S_I 2.48 S_R 0.95 U_o	U_L	V_F	W_E 15 m S_I S_R U_o	U_L	V_F
20	0.65	0.82	26	0.65	0.82	25	0.65	0.82	24	0.64	0.82	24			
22	0.65	0.82	24	0.65	0.81	22	0.65	0.81	23	0.65	0.81	22			
24	0.66	0.83	23	0.65	0.82	22	0.65	0.83	21	0.66	0.83	21			
26	0.63	0.74	21	0.62	0.74	21	0.61	0.74	21	0.61	0.72	28			
28	0.62	0.70	21	0.62	0.70	20	0.60	0.69	19	0.60	0.69	18			
30	0.58	0.70	19	0.58	0.69	19	0.58	0.69	18	0.57	0.68	18			
32	0.55	0.62	19	0.53	0.62	18	0.51	0.61	18	0.50	0.62	17			
34	0.52	0.58	18	0.51	0.57	17	0.48	0.56	17	0.48	0.56	16			
36	0.46	0.53	17	0.45	0.54	16	0.44	0.54	16	0.44	0.53	16			
38	0.45	0.51	16	0.43	0.50	16	0.41	0.49	16	0.40	0.49	16			
40	0.44	0.49	16	0.42	0.48	16	0.40	0.47	15	0.40	0.45	15			
42	0.45	0.47	16	0.43	0.46	15	0.42	0.45	15	0.41	0.44	15			
44	0.44	0.46	15	0.43	0.45	15	0.42	0.44	14	0.40	0.43	14			
46	0.44	0.47	14	0.43	0.45	14	0.40	0.44	14	0.40	0.44	14			
48	0.43	0.44	14	0.40	0.43	14	0.37	0.42	14	0.37	0.42	13			
50	0.42	0.39	14	0.39	0.38	13	0.36	0.38	13	0.34	0.38	13			

Table 10.3   A selection of luminaire maintenance factors (LMFs)

Cleaning interval (months)	Ingress protection of lamp housing					
	IP2[a] minimum (see BS 5490)			IP5[a] minimum (see BS 5490)		
	Pollution category			Pollution category		
	High	Medium	Low	High	Medium	Low
12	0.53	0.62	0.82	0.89	0.90	0.92
18	0.48	0.58	0.80	0.87	0.88	0.91
24	0.45	0.56	0.79	0.84	0.86	0.90

[a] The second number (see Table 5.4) is not needed.

Table 10.4   A slection of approximate lamp flux depreciation factors (LFDFs)

Lamp wattage and type	Replacement period (h)				
	4000	6000 (2 years)	8000	10 000	12 000 (4 years)
125 W MBF	0.92	0.88	0.85	0.83	0.80
400 W MBF	0.92	0.89	0.88	0.85	0.84
55 W SOX	0.97	0.96	0.94	0.92	0.91
131 W SOX	0.99	0.99	0.98	0.97	0.96
70 W SON-T	0.95	0.93	0.91	0.88	0.85
150 W SON-T	0.99	0.98	0.97	0.96	0.95
250 W SON-T	0.98	0.97	0.96	0.94	0.92

> pollution factor – medium
> cleaning interval – 12 months
> lamp replacement – 10 000 h

Referring to Table 10.3, the LMF is 0.90 and from Table 10.4, the LFDF is 0.96. Hence the MF is 0.90 × 0.96 which is 0.86. Therefore,

$$S = \frac{3.38 \times 16.0 \times 0.86}{1}$$

$$= 46.5 \text{ m}$$

Referring back to Table 10.2 for S of 46 m, W of 8 m,

$U_o = 0.50$
$U_L = 0.50$
$V_F = 15$

Referring back to Table 10.1, U_o and U_L meet the uniformity requirements. V_F is the 'veil factor' which can be used to calculate the threshold increment TI. However, it is normally reasonable to accept the manufacturer's classification.

The result of these calculations indicates a satisfactory layout. There are a number of permutations of mounting height (H), lantern type (MTI or LTI), lamp flux (ϕ) and layout. The final scheme is a result of cost, the designer's experience, and aesthetics.

Self-assessment task 10.2

Calculate the spacing (S) if the lanterns are only cleaned once every 2 years.

Spacing on bends

The patterns of luminance patches interlock on a straight stretch of road. Even a gentle bend in plan, when seen by the driver, cuts off all but the close patches. To maintain even luminance it may be necessary to reduce the spacing S and only site the columns on the outside of the bends.

To help design the layout, a transparent template is made up as shown in Fig. 10.7. This shows the isoluminance contours of 12.5 per cent maximum (A) and 25 per cent (C). It should be drawn to the same scale as the road plan, e.g. 1:500, and the small circle at the bottom represents a driver 90 m from the lantern. The template will only apply to a specific H, in this case 10 m.

Use of the template is shown in Fig. 10.8. It indicates the patch of luminance that the driver will see from various points around the bend. The shaded area indicates a possible dark patch and the spacing S needs to be reduced. The patterns should be checked for both directions (except for a dual carriageway).

Curve B is the 12.5 per cent contour for two lanterns back to back.

Road junctions

There is a wide variety of junctions and these are dealt with in BS 5489 Part 4. Junctions are the most likely places for accidents and the lighting must:

- Warn the driver of an approaching junction.
- Reveal any traffic emerging from minor roads.
- Not provide the additional hazard of columns sited at the junction blocking the view.

Figures 10.9–10.12 show typical layouts for:

- A T junction   Fig. 10.9
- A crossroad   Fig. 10.10
- A mini-roundabout   Fig. 10.11
- A roundabout   Fig. 10.12

The principle is to maintain even luminance, site a column beyond an intersection (no more than $\frac{1}{3}S$), and, where possible, keep to the outside of bends. When approaching a major road from a minor one try to site a column on the far side of the intersection. It all makes sense if one can anticipate what drivers need to see, whether they have to halt or be aware of any emerging traffic.

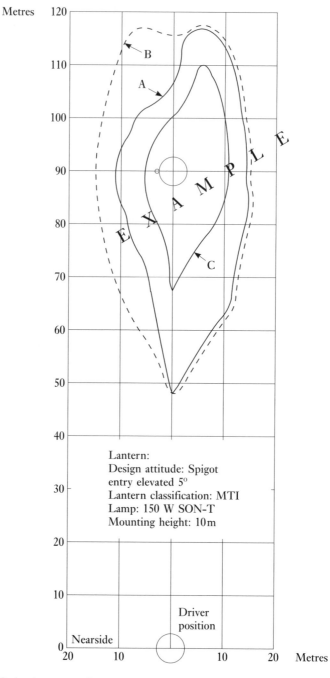

Fig. 10.7   Isoluminance templates

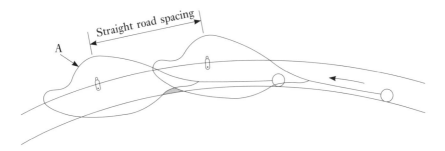

Fig. 10.8 Inadequate coverage of road

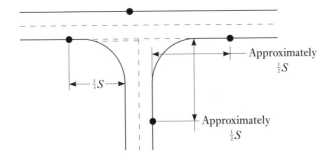

Fig. 10.9 T junction (*S* is the design spacing for the major road)

Pedestrian crossings

Similar principles apply to pedestrian crossings and Fig. 10.13 shows requirements for staggered and opposite layouts. The layout becomes more complex when crossings are at road junctions. It is to be hoped that the junction layout is satisfactory for the crossing.

Part 3: *Lighting for subsidiary roads and associated pedestrian areas*

Main traffic routes consider the needs of the driver, and everything else, pedestrians included, is a hazard to be seen and avoided. Many roads with lower traffic speeds and density are more social areas where the aims are not so heavily driver orientated. The pedestrian is more concerned with illumination and seeing in a similar fashion to an interior situation.

The recommendations are given in the form of average and minimum horizontal illuminances. There are three categories of road which are graded according to pedestrian and traffic use, and crime risk (Table 10.5).

There is a limited degree of glare control. This applies only to lanterns emitting more than 3500 lm in the lower hemisphere. In this case the maximum acceptable intensity at 80° is 160 cd/klm and at 90° is 80 cd/klm.

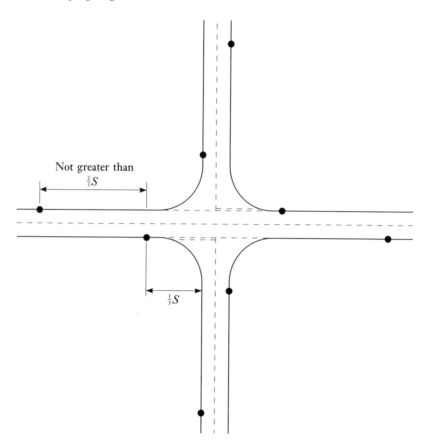

Fig. 10.10   Crossroad (*S* is the design spacing for the major road)

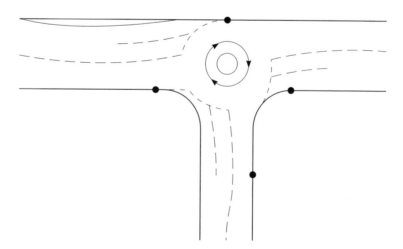

Fig. 10.11   Mini-roundabout

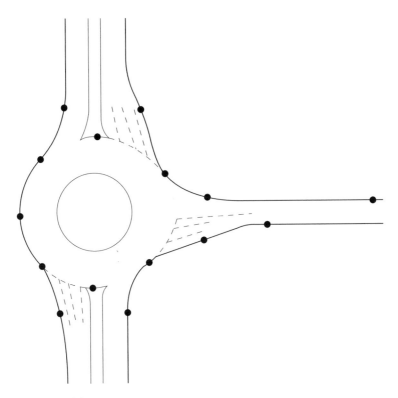

Fig. 10.12 Roundabout

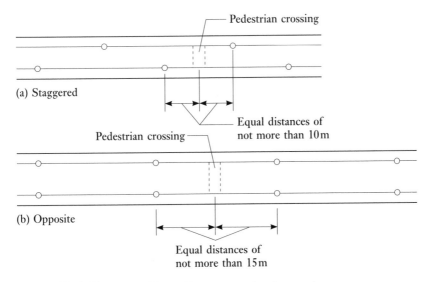

Fig. 10.13 Typical lantern positions adjacent to a pedestrian crossing

Table 10.5   Illuminance levels

Category	Average illuminance	Minimum illuminance
3/1	10	5
3/2	6	2.5
3/3	3.5	1

Category 3/1   Roads where night-time public use is likely to be high, or the crime rate is likely to be high, or traffic usage is likely to be high

Category 3/2   Roads that do not fall into category 3/1 and where night-time public use is likely to be moderate, or the crime rate is average to low, or traffic usage is of a level equivalent to that of a housing estate access road

Category 3/3   Roads where night-time public use is minor and solely associated with the adjacent properties, and the crime rate is very low, and traffic usage is of a level equivalent to that of a residential road

Planning the layout

The steps are:

1. Decide on a type of layout (e.g. staggered), mounting height H, lantern type and lamp.
2. Calculate the spacing S to achieve the minimum illuminance.
3. Using the value of S calculate the average illuminance to see if it complies with the category.
4. Site the columns without exceeding S.

Example

A road has a width of 10 m and a single pathway of 3 m. Lanterns with 70 W SON lamps are to be mounted on 6 m columns and the photometric data are as given in Fig. 10.14. The area is designed category 3/2. A single-sided layout is to be used with the lanterns over the kerb (no overhang). Calculate (a) the minimum illuminance, (b) the average illuminance area, (c) the glare.

Solution (a)

There are various places where minimum illuminance can occur. To simplify the calculation, it will be assumed that it occurs in the road centre, midway between columns. A spacing S has to be assumed and this will be taken as 33 m.

An isolux diagram is given for an H of 1 m (see Fig. 10.14). This can be used to find the minimum illuminance.

If H is 6 m and S is 33 m, then midway $S/2$ will be 16.5 m. Expressing this in terms of H,

$$S/2 = \frac{16.5}{6} H$$

$$= 2.75H$$

Description: Road lighting lantern
Recommended column height(s): 5–6m

Version	Catalogue number	Lamp
–	QB5 1050.4	50 W SON–E
–	QB5 1070.4	70 W SON–E

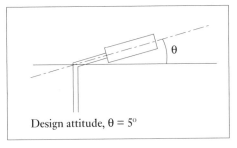

Design attitude, $\theta = 5°$

Isolux diagram lx/klm for $H = 1$m, lantern at (0,0)

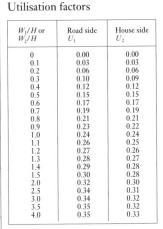

Utilisation factors

W_1/H or W_2/H	Road side U_1	House side U_2
0	0.00	0.00
0.1	0.03	0.03
0.2	0.06	0.06
0.3	0.10	0.09
0.4	0.12	0.12
0.5	0.15	0.15
0.6	0.17	0.17
0.7	0.19	0.19
0.8	0.21	0.21
0.9	0.23	0.22
1.0	0.24	0.24
1.1	0.26	0.25
1.2	0.27	0.26
1.3	0.28	0.27
1.4	0.29	0.28
1.5	0.30	0.28
2.0	0.32	0.30
2.5	0.34	0.31
3.0	0.34	0.32
3.5	0.35	0.32
4.0	0.35	0.33

Glare control data

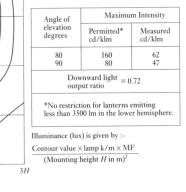

Angle of elevation degrees	Maximum Intensity	
	Permitted* cd/klm	Measured cd/klm
80	160	62
90	80	47
Downward light output ratio	= 0.72	

*No restriction for lanterns emitting less than 3500 lm in the lower hemisphere.

Illuminance (lux) is given by :–

$$\text{Contour value} \times \text{lamp k/m} \times \text{MF} \over (\text{Mounting height } H \text{ in m})^2$$

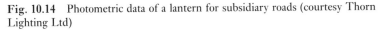

Fig. 10.14 Photometric data of a lantern for subsidiary roads (courtesy Thorn Lighting Ltd)

The distance to the road centre will be

$$W/2 = \frac{5}{6}H$$

$$= 0.83H$$

This point P can be located on the isolux diagram and gives a value of 5 lx, which now needs converting for the actual values of H, ϕ and MF.

ϕ gives the initial lumens and can be found from the lamp data in Appendix C, i.e. 5.8 klm.

MF is found as before from Tables 10.3 and 10.4. Assuming IP5 and medium pollution with a 12 month cleaning interval, LMF is 0.90.

The LFDF on an 8000 h replacement is 0.91. The MF is 0.9×0.91, i.e. 0.82. Therefore,

$$E = \frac{\text{isolux value} \times \phi \times \text{MF}}{H^2}$$

$$= \frac{5 \times 5.8 \times 0.82}{36} \qquad [10.2]$$

$$= 0.66 \text{ lx}$$

This must be doubled to allow for the contribution from the next lantern, so

$$E = 2 \times 0.66$$

$$= 1.3 \text{ lx minimum}$$

Reference back to Table 10.5 will show this value to be too low – the minimum should be at least 2.5 lux (double the value). This means 10 lx is needed on the isolux diagram. This occurs at Q when $S/2$ is $2.3H$ or 14 m. The design spacing S can be 28 m.

Solution (b)

$$E = \frac{1000(U_1 + U_2) \times \phi \times \text{MF}}{(W_1 + W_2)S} \qquad [10.3]$$

where

E	is the average horizontal illuminance (in lx)
U_1	is the road side UF
U_2	is the house side UF
ϕ	is the lamp flux (in klm)
MF	is the maintenance factor
W_1	is the road side width
W_2	is the house side width
S	is the lantern spacing (in m)

Values for U_1 and U_2 are found from Fig. 10.14:

$$H = 6 \text{ m}$$
$$W_1 = 10 \text{ m}$$
$$W_2 = 3 \text{ m}$$
$$\therefore \quad W_1/H = 1.66$$
$$W_2/H = 0.5$$

From Fig. 10.14,

$$U_1 = 0.31$$
$$U_2 = 0.15$$
$$\phi = 5.8 \text{ klm}$$
$$MF = 0.82$$
$$E = \frac{1000 \times (0.31 + 0.15) \times 5.8 \times 0.82}{(10 + 3) \times 28}$$
$$= 6.0 \text{ lx}$$

This could be said to comply, just, for category 3/2 (6 lx).

Self-assessment task 10.3

Repeat the calculations but with the lanterns mounted 5 m above the road.
   Are there any advantages in reducing the mounting height?

Solution (c)
The downward lumens are given by

$$\phi \times \text{DLOR} = 5.8 \times 0.72$$
$$= 4.17 \text{ klm}$$

As this is above 3.5 klm, the intensity values have to be checked. The photometric data in Fig. 10.14 give these values and they are satisfactory. In fact a responsible manufacturer is unlikely to produce lanterns which do not comply.

Outdoor light pollution

There is no merit in lighting the sky and both road lighting and floodlighting should be designed to avoid the upward spread of light called 'Sky glow'. Complete elimination is impossible as light will be reflected upwards from roads and ' buildings, but the system of lighting can be designed so that light only falls on the surfaces to be illuminated.

   The Institution of Lighting Engineers (ILE) has published *Guidance notes for the reduction of light pollution 1994* and its recommendations are set out here in full.

   Listed below are some easy ways to reduce the problems of unnecessary obtrusive light:

(A1) Switch off lights when not required for safety, security or enhancement of the night-time scene. In this respect one can introduce the concept of a curfew with further limitations on lighting levels between agreed hours, e.g. advertising and decorative floodlighting – off between 11 p.m. and dawn.
(A2) Direct light downwards wherever possible to illuminate its target, not upwards. If there is no alternative to uplighting, then the use of shields and baffles will help reduce spill light to a minimum.
(A3) Use specifically designed lighting equipment that once installed minimises the spread of light near to or above the horizontal.
(A4) Do not 'over' light. It is a cause of light pollution and a waste of money.

(A5) To keep glare to a minumum, ensure that the main beam angle of all lights directed towards any potential observer is kept below 70°. It should be noted that the higher the mounting height, the lower can be the main beam angle. In places with low ambient light, glare can be very obtrusive and extra care should be taken in positioning and aiming.

(A6) Wherever possible use floodlights with asymmetric beams that permit the front glazing to be kept at or near parallel to the surface being lit.

(A7) For domestic and small-scale security lighting, there are two solutions:
   (i) Passive infrared detectors can be used to good effect, if correctly aligned and installed. A 150 W (2000 lumen) tungsten halogen lamp is more than adequate; 300/500 W lamps create too much light, more glare and darker shadows.
   (ii) All-night lighting at low brightness is equally acceptable. For a porch light a 9 W (600 lumen) compact fluorescent lamp is adequate in most locations.

(A8) For road lighting, light near to and above the horizontal should be minimised. (Note UWLRs in Table 10.6) If you are a lighting engineer, planning authority and/or involved in the design or installation of lighting, the ILE strongly recommends the adoption of the following obtrusive light limitations for exterior lighting installations (Table 10.6).

Environmental zones   It is anticipated that the following zones will be defined by the local planning authority:

E1:   In 'National Parks', 'Areas of outstanding natural beauty' or other 'Dark landscapes'.
E2:   Areas of 'low district brightness' (e.g. in a rural location, but outside an E1 zone).
E3:   Areas of 'medium district brightness' (e.g. in an urban location).
E4:   Areas of 'high district brightness' (e.g. in an urban centre with high night-time activity).

Table 10.6   Obtrusive light limitations for exterior lighting installations

Environmental zone	Sky glow UWLR (max %)	Light into windows E_V (lx) Before curfew	Light into windows E_V (lx) After curfew	Source intensity I (kcd) Before curfew[b]	Source intensity I (kcd) After curfew	Building luminance[c] L (cd/m²) average before curfew
E1	0	2	1[a]	0	0	0
E2	5	5	1	50	0.5	5
E3	15	10	5	100	1.0	10
E4	25	25	10	100	2.5	25

UWLR (Upward Waste Light Ratio) is the maximum permitted percentage of luminaire flux that goes directly into the sky, E_V is the vertical illuminance in lux, I the light intensity in candelas and L the luminance in candelas per square metre.

[a] Acceptable from public road lighting installations *only*.

[b] This applies to each source in the potentially obtrusive direction, *outside* of the area being lit. The figures given are for general guidance only and for some medium to large sports lighting applications with limited mounting heights may be difficult to achieve. However, if the aforementioned recommendations are followed then it should be possible to lower these figures to under 10 kcd.

[c] This should be limited to avoid overlighting, and relate to the general district brightness.

CONCLUSION

The preceding pages have explained the current techniques for designing the lighting of roads. Much of this can now be done with suitable computer programs. But, as with all lighting design, there is a range of options in choice of equipment, layout and geometry. If this were not so, we would never need to remark 'well, that's a well-lit road'. They would all be the same!

EXERCISES

10.1   Explain the principle of silhouette vision and how it can be applied to the lighting of the following:
(a) a crossroad
(b) a T-junction
10.2   Discuss the differences in principle between the lighting of main roads and subsidiary roads.

Using the data in Fig. 10.14 calculate the illumination 10 m down the road using a 50 W SON lamp on a 5 m column for a new lamp in a clean lantern. The point of measurement is in the centre of a road 12 m wide.

CHAPTER 11

FLOODLIGHTING

TOPICS COVERED

FLOODLIGHTING A BUILDING
Planning
Illuminance levels
Lumen calculation

FLOODLIGHTING A HORIZONTAL OPEN AREA
Planning
Sources of specialised data

Floodlighting, as the term implies, is literally flooding a surface with light. If this is all that is needed, then the design and calculations are based on achieving an acceptable level of illumination and uniformity on vertical surfaces, e.g. buildings, or on horizontal surfaces, e.g. car parks. This chapter will deal with these calculations.

There is often much more to outdoor lighting, such as:

- appearance during daytime
- glare from the installation
- decorative lighting
- lighting for specific outdoor activities such as sport

A list of relevant publications which give guidance on these aspects is included at the end of the chapter.

FLOODLIGHTING A BUILDING

The general principles of illuminance calculations have been covered in earlier chapters. A practical example will now be taken to explain the design process.

Planning

Figure 11.1 shows a church. There is plenty of space available for siting the floodlights and even lighting is required on the whole elevation.

The basic requirements are the provision of even and sufficient illuminance. Successful floodlighting also requires a sense of drama and colour. In general, all floodlighting of a flat plain surface should be reasonably uniform. There is a greater degree of tolerance than for an interior lighting scheme, but all the beams from each floodlight should overlap. A suggested uniformity ratio (maximum:average) is 5:1.

Draw a series of overlapping circles to cover the whole elevation with the minimum of waste light. This is shown in Fig. 11.2.

A circle represents a beam of light whose edge is about one-tenth the illuminance of the centre; these circles should be drawn so that the edge of one passes through the centre of the next beam. Provided that the floodlights are a reasonable distance away and have symmetrical beams this technique will produce an even floodlighting effect.

In this example, nine floodlights will provide coverage of the church and tower. If the distance from the floodlights to the church is 50 m, and as each beam will be about 14 m wide, the beam width will be given by

$$\tan\frac{\text{beam angle}}{2} = \frac{\text{radius of beam}}{\text{distance}}$$
[11.1]

and in this case

Fig. 11.1

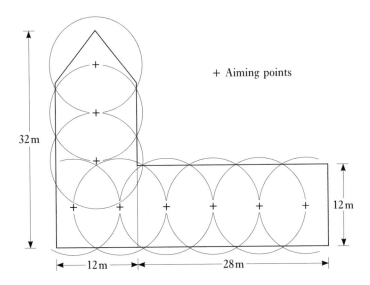

32 m

+ Aiming points

12 m

12 m

28 m

Fig. 11.2

Table 11.1   Beam data

Floodlight	Peak intensity (cd/1000 lm)	Beam efficiency (% total beam)	Beam angles			
			10% peak		50% peak	
			Vertical	Horizontal	Vertical	Horizontal
NNF 020/1 400 W SON Narrow beam	1455	68	$2 \times 37°$	$2 \times 37°$	$2 \times 15°$	$2 \times 15°$
NNF 020/2 400 W SON Wide beam	1416	76	$2 \times 30°$	$2 \times 39°$	$2 \times 17°$	$2 \times 17°$
NNF 020/1 250 W SON-T Narrow beam	20 500	60	$2 \times 7°$	$2 \times 7°$	$2 \times 2°$	$2 \times 2°$
NNF 020/2 250 W SON-T Wide beam	7585	62	$2 \times 14°$	$2 \times 14°$	$2 \times 4°$	$2 \times 4°$

Source: Philips Lighting Ltd.

$$\tan\frac{\text{(beam angle)}}{2} = \frac{7}{50}$$
$$\therefore \quad \text{beam angle} = 2\tan^{-1}(7/50)$$
$$= 16°$$

Floodlights can be categorised by their beam angles, e.g. narrow beam, but these classifications are sometimes vague and it is better to work with actual beam angles, as these are normally quoted by the supplier.

A typical set of floodlighting data is given in Table 11.1. This is from Philips Catalogue, NNF 020 Floodlight range, and has some narrow beam types. NNF 020/1 has a 14° beam ($2 \times 7°$) and a 250 W SON-T lamp gives 28 500 lm and looks a possible solution.

At this stage it is worth checking the illuminance at the beam peak (the maximum). It is 20 500 cd/1000 lm. Therefore,

$$\text{Actual intensity} = 20\ 500 \times \frac{28\ 500}{1000}$$
$$= 584\ 250\ \text{cd}$$
$$\therefore \quad \text{Peak (or max) illuminance} = \frac{584\ 250}{\text{(distance)}^2}$$
$$= \frac{584\ 250}{50^2}$$
$$= 234\ \text{lx}$$

Table 11.2 Recommended building illuminances for different district brightnesses

Building condition	High	Medium	Low
Clean	90	55	35
Fairly clean	140	90	50
Dirty	280	180	100

Notes: High – town centre
Medium – suburban
Low – Rural.
Source: Philips Lighting Ltd.

Applying a maintenance factor of 0.7,

Peak illuminance $= 234 \times 0.7$
$= 164 \text{ lx}$

Illuminance levels

The amount of light depends on a number of factors. The basic requirement is that the floodlit area will stand out from its surroundings. Table 11.2 gives a guide on illuminance values, assuming a building reflectance of 0.35 (medium stone).

If, in this example, the church is on the outskirts of a town and is made of stone in a fair state of cleanliness, the peak illuminance of 164 lx is a reasonable design level.

Lumen calculation

To check the average illuminance a basic lumen calculation is used:

$$\text{Average illuminance} = \frac{N \times L \times \text{MF} \times \text{BF} \times \text{WLF}}{\text{total surface data}} \qquad [11.2]$$

where

N	is the number of floodlights
L	is the initial lamp lumens per floodlight
MF	is the maintenance factor
BF	is the beam lumens/lamp lumens (beam efficiency)
WLF	is the waste light factor; some of the beam will miss the building and this figure is taken as about 0.7 of the light reaching its target

$$\therefore \quad E = \frac{9 \times 28\,500 \times 0.7 \times 0.6 \times 0.7}{(12 \times 32) + (12 \times 28)}$$

$$= 122 \text{ lx}$$

Again this value would appear to be satisfactory. As already stated, this is just part of designing a floodlighting scheme. The following is a list of other considerations:

Modelling – the three-dimensional shape and architectural features should be revealed by suitable direction of the floodlights.

Colour – if feasible, the colour of the lamp should be different to those used in the general environment.

Glare – beams must be concealed from the principal directions of view.

Sky glow – see page 232.

Self-assessment task 11.1

Will the church floodlighting scheme meet the light pollution requirements in Chapter 10 if the church is in an urban location?

FLOODLIGHTING A HORIZONTAL OPEN AREA

The process for lighting a horizontal area is similar to that used for the church, but it is more accurate to use horizontal illuminance plots such as isolux diagrams (see Chapter 2). In the next example the requirement is to illuminate a large car park using floodlights on 8 m high poles. An illuminance of 20 lx is required (CIBSE recommends 5–20 lx, 50 lx for high-risk areas; refer to Table 11.3).

Planning

Figure 11.3 gives the isolux data for a 400 W MBF floodlight on a flat plane 8 m below the floodlight. This is the distribution of illuminance on a plane parallel to the front of the floodlight and normal to the axis of the beam. The broken line on the figure shows the boundary where the illuminance has fallen to one-tenth of the maximum, i.e. this is the approximate area that a single floodlight would cover. The average illuminance would not be the maximum but would lie somewhere between the maximum and boundary value.

Table 11.3   Recommended illuminance levels for outdoor areas

Location	Category	Average horiz. lx level
Car park	High risk	50[a]
Car park	Low risk	20[a]
Pedestrian precinct	High risk	100[a]
Pedestrian precinct	Low risk	20[a]
Service area		50[a]
Storage area		20[a]
Tennis	Club	300[b]
Football	Club	140[b]
Football	Recreational	50[b]
Athletics	Club	200[b]

[a] From CIBSE Guide 1994.
[b] From CIBSE Sports Guide 1990.

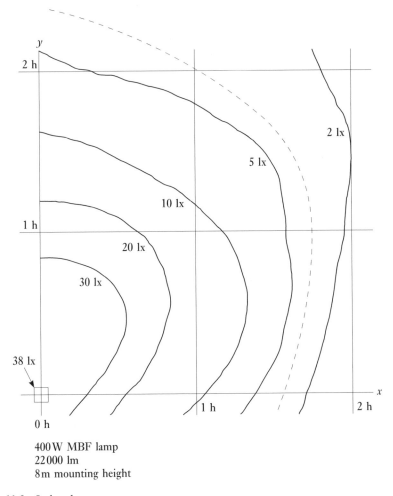

400 W MBF lamp
22 000 lm
8 m mounting height

Fig. 11.3   Isolux data

It is important to appreciate that the isolux diagrams refer to a specific distance and direction. They can, however, be used for other distances and planes at different angles. This involves the use of the inverse square law and cosine law. If the distance is increased, then the illuminance will decrease but the areas will increase correspondingly.

A simplified method of design can be set out as follows:

1. Choose the most suitable type of lamp and wattage for the application.
2. Choose the most suitable type of floodlight:
   400 W MBF lamp
   22 000 lm (see Appendix C)
   8 m mounting height

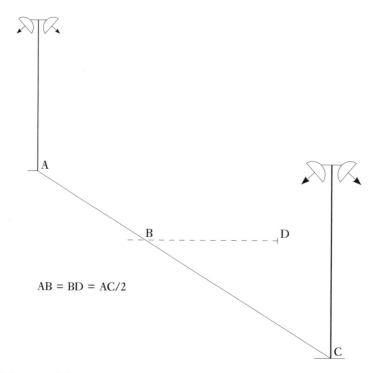

AB = BD = AC/2

Fig. 11.4   Example layout

3. Choose the mounting height to be used.
4. Determine the maximum spacing to achieve an acceptable diversity.
5. Check that the scheme will provide the design level of average illuminance.

In this case the choice is between a cluster of floodlights at the top of each pole, or a single post-top lantern (street lighting lanterns are not satisfactory, as their light distribution is not designed for this situation). High-wattage lamps can be used and the initial design is two 400 W MBF floodlights mounted back to back. Their isolux diagram for 8 m mounting is shown in Fig. 11.3 and a diversity in illuminance of 10:1 should be acceptable.

Using a layout as illustrated in Fig. 11.4 the illuminance directly below each pair of floodlights will be 76 lx. At point B midway between two poles the illuminance needs to be at least 7.6 lx for a diversity of 10:1. To achieve this, each floodlight must provide 7.6/4 lx which will occur when AB is approximately 15 m. Hence the spacing AC will be 30 m.

Next consider the distance along the car park. With 30 m spacing, check the illuminance at the midpoint D. This is 2 lx from each floodlight, giving a total of 8 lx. This is satisfactory (it exceeds the 7.6 lx requirement).

The average illuminance can be checked using the lumen method. Each floodlight effectively lights an area 30 m wide × 15 m across, i.e. 450 m². Using eqn [11.2] and assuming the same values for MF, BF and WLF,

$$\text{Average illuminance} = \frac{22\,000 \times 0.7 \times 0.6 \times 0.7}{450}$$

$$= 14 \text{ lx}$$

This is below the design requirements of 20 lx. In practice, owing to the diversity of 10:1 it is doubtful if the design would be altered. It is unrealistic to be too precise when assumptions, like the factors, are made.

To get a 20 lx average illuminance the poles would have to be spaced closer or, if this is unsatisfactory, then another floodlight would have to be considered. The area to be covered by each floodlight would be reduced to 450 × 15/20 or 337 m². Another solution could be to use 400 W SON PLUS lamps, if suitable. This would raise the illuminance by 54 000/22 000 to 34 lx. This assumes that the floodlight will take either type of lamp.

Unfortunately suppliers of floodlighting equipment may not provide isolux diagrams in their technical data. The choice then is either to draw a diagram from the intensity distribution data, or to ask the photometric department of the supplier for the diagram.

Self-assessment task 11.2

Is there an alternative mounting height that could be more economical for lighting this car park?

Sources of specialised data

Building floodlighting

Electricity Council, *Outdoor Lighting*, 1983
CIBSE, *The Outdoor Environment*, 1992

Outdoor areas

CIBSE Guides: *Building and Civil Engineering Sites*, 1975
Shipbuilding and Ship Repair, 1979
The Industrial Environment, 1989

Outdoor sports

CIBSE Guide: *Sports*, 1990

EXERCISES

11.1   An advertising hoarding is 5 m wide by 3 m high with 3 m clear space in front. It is to be illuminated to 100 lx with a uniformity of 0.5. The site is in a town centre. Design a suitable floodlighting scheme using the data in Table 11.1.

11.2   A club tennis court, 11 m × 23.8 m, is to be lit from the sides by floodlights on 6 m poles. Using the data in Fig. 11.3 design a scheme to achieve a uniformity of 0.5.

COMPUTER PROGRAMS

This appendix discusses the advantages and limitations of using computer-aided design (CAD) in the preparation of lighting layouts. Some of the available programs are outlined for use in calculations, visualisation and information technology.

INTRODUCTION

The computer is a tool which can help the designer and photometrician carry out repetitive calculations at speed. In earlier days they used log tables and slide rules. These did not design schemes for them but reduced and simplified the calculations. CAD is a 'giant step forward' but it is still only a tool requiring an intelligent operator. One marvels at what it can do but the designer needs to understand what it is doing.

The purpose of this book is to educate the reader in the science and application of light. The reader can use any computing aid available, but the level of lighting knowledge must be sufficient to understand the limitations of CAD when applied to lighting.

HOW CAN CAD HELP?

Many calculations are fairly basic but repetitive. For example, football pitch floodlighting is usually planned using CAD. There are specific recommendations on illuminance levels and uniformity. The location of the floodlights will be either on towers or the roof of the stands. The computer can supply solutions in minutes that would normally take days.

However, a decorative interior such as a restaurant or theatre requires decisions on appearance and visual interest which could baffle the computer. Visualisation programs can provide a guide to the relative appearance of room surfaces under different forms of lighting but they do require a high degree of data input involving time and expense, neither of which is normally available.

Daylight calculations are suitable for CAD and a number of programs are available. They enable quick decisions to be made on window size and position. An important difference between electric lighting and daylighting programs is that most electric lighting requires the manufacturer's own data input and may not be of use in comparing equipment performance from competing companies. For daylighting the window is a flat opening and although different types of glass may be considered, their performance is not going to vary in a significant way with different glass manufacturers.

EXAMPLES OF COMPUTER PACKAGES

There is a fairly wide range of software available and this chapter restricts itself to three examples that are in use today.

'Calculux for Windows' – Philips Lighting

There are three software programs:

1. Calculux Indoors: this will perform illuminance calculations including direct, indirect and average values for orthogonal rooms. It requires the user to enter a maintenance factor. It will also calculate the glare index and carry out cost analysis.
2. Calculux Area: this will plan outdoor floodlighting schemes. It can calculate illuminances on horizontal, vertical and semi-spherical planes. It can also calculate veiling luminance and glare rating. As with 1 it can carry out cost analysis.
3. Calculux Road: this will calculate spacing for a straight stretch of road for the different road classification. Again it will carry out a cost analysis. It will not help on the more difficult problems such as bends and road junctions.

All these programs are on 3.5 inch floppy disks and the packages are complete with instructions and database. All three programs are based on Philips lamps and luminaires, but the facility exists for storing data from other manufacturers.

For more information contact: Philips Lighting, 420–430 London Road, Croydon CR9 3QR. Tel.: 0181 665 6655, fax: 0181 684 0136.

'Lighting Visualisation' – Whitecroft Lighting

Lighting calculations provide fundamental technical information but they do not demonstrate what a scheme will look like. Programs are available which give an indication, if not a completely accurate picture, of a lit interior. The technique used is known as 'ray tracing' and the computer literally follows the path of a ray of light from the luminaire to the room and furniture surfaces and then back (by reflection) to the viewpoint. An example, Fig. A.1, is shown of a three-dimensional visualisation output from PowerPlan (Whitecroft Lighting), where furniture and

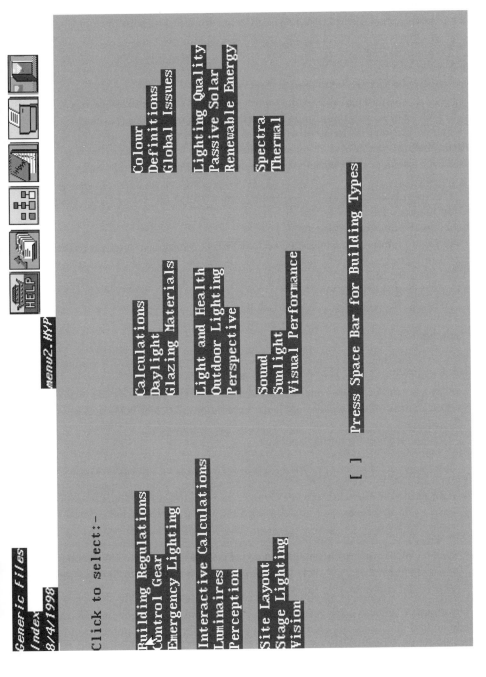

Fig. A.2 Example of Hyperlight text. © HyperLight Partnership.

objects can be included. All of the standard illuminance data can be produced, but the program can also derive simple visualisations from these results, including shadow effects.

Commercial packages can go further by allowing the user to assign materials and reflectivities to surfaces such as marble or wood. The program can render the view to increase the accuracy of the image. Successive images can then be rendered to generate a 'walkthrough' as though the user is moving through the room, often called VR (Virtual Reality). It must be remembered, however, that the more complex a program the longer it will take to learn to use, to set up and to calculate (depending on complexity). A computer such as a 200 MHz Pentium (Intel) can take days to render and calculate owing to the huge number of calculations required.

One must balance the effort required in setting up the program with the desired result to satisfy the customer. As the technology improves, the results will get better and be quicker to calculate.

For more information contact: CD-ROM Project Manager, Whitecroft Lighting Ltd, Burlington Street, Ashton-under-Lyne OL7 0AX. Tel: 0990 087 087. Or: Transformation Software Ltd, Visual Simulation & Virtual Reality Division, Thame Park Road, Thame OX9 3UQ. Tel: 01844 261456.

Hyperlight 3

This is the lighting *Encyclopaedia Britannica* in that its aim is to provide computer access to all general aspects of lighting. By the use of 'hypertext', a lighting topic, e.g. daylight, can be entered and, by following the screen instructions and choices, the development of this topic in particular applications, e.g. roof lights, can be followed.

To quote from the company's publicity:

Information is divided into 'frames', each small enough to fit on a computer screen. On the screen, or within each frame, about half a dozen key words or phrases might be highlighted. The user selects one key word at a time for expansion. This places another frame on the screen, elucidating that key word and offering a choice of half-a-dozen further key words to investigate.

Every user will make a different journey through the information. One may feel a need to probe systematically deeper and deeper into the fundamentals. Others may wander from frame to frame, as successive highlights whet their appetite for additional tit-bits.

The program will also carry out illuminance calculations for daylight and electric light and, in its latest form, includes the Hunt model for colour appearance and the Rea model for visual performance. Figure A.2 shows a typical menu display indicating a choice of subjects to be investigated.

For more information contact: The HyperLight Partnership, 4 Aigburth Avenue, St Georges Road, Hull HU3 3QA. Tel/fax: 01482 216792.

As stated, these are just examples of a wide range of programs in use by the majority of technical departments in the lighting manufacture and design industry. There is no single centre of information but it may well be worth contacting: Building Services Centre, Building Research Establishment, Garston, Watford WD2 7JR. Tel: 0171 505 6622 (Publications).

LEGAL REQUIREMENTS

Lighting is seldom a matter of legal action. It could be a factor in a road accident at night, but it may be difficult to cite it as a possible cause with interior lighting, except, perhaps, where there is no light at all.

If lighting is designed to the recommendations of the CIBSE and appropriate British and European standards it will satisfy all legal requirements. It is important to realise that the law is concerned with the safety of its subjects. The codes are concerned both with safety and with efficient visual conditions and the visual environment. The latter will normally require higher lighting standards than the former. It is, however, useful to have knowledge of relevant Statutory Instruments (SIs) and EEC Directives.

UK legislation

Health and Safety at Work Act 1974

Sections 2 and 4 state that employers should provide lighting appropriate to the work space and its activities.

Offices, Shops, and Railway Premises Act 1963

This refers to 'suitable and sufficient' for both daylight and electric lighting.

EEC Directives relevant to lighting

89/654

This is concerned with minimum safety and health requirements in the workplace. The CIBSE recommendation of not less than 200 lx in all continuously occupied work areas more than meets any standards required for safety. The UK SI is 3004.

250

90/270

This is concerned with requirements for work with display screen equipment. For lighting it uses general statements such as 'no disturbing reflections in screens'. CIBSE LG3 – see Chapter 8 – provides specific guidance on meeting EEC requirements. The UK SI is 2792.

92/58

This is concerned with emergency lighting, the design of signs, and the use of colour to indicate hazards.

Electric lighting requirements of Building Regulations Part L 1995

These are requirements on the use of energy-efficient lighting for all new non-domestic buildings of over 100 m² floor area, and to certain existing buildings undergoing change of use. To meet the requirements:

1. Average lamp circuit efficacy should be over 50 lumens/watt.
2. Flexible lighting control should enable effective use to be made of daylighting.
3. Energy efficiency and controls, as recommended in the CIBSE 1994 Code for interior lighting, should be provided.

The Regulations contain a list of recommended lamps which include all discharge lamps with the exception of compact fluorescent lamps below 11 watts and mercury-blended lamps.

The lamp circuit efficacy is

$$\frac{\text{Initial lumen output of lamp after 100 hours}}{\text{lamp watts} + \text{control circuit watts}}$$

Lamps which do not comply may be used, e.g. tungsten filament, provided 95 per cent of the lighting load comprises high-efficacy lamps.

Problems can arise in areas used principally for display, e.g. shops, museums, restaurants. This then becomes a matter for discussion between the designer and building control officer. The inference is that a case can be made for the use of lower-efficacy lamps where there are good grounds for using that particular lighting system.

It is important to appreciate that the requirements apply to the whole building rather than each room. Examples of how to apply the Regulations are given in the CIBSE Guidance Note GN4 1996.

Legislation for daylight

Rights of light

Common law rules, statutory rules, and procedures that govern the acquisition of the right of access to and enjoyment of light received over neighbouring land are complex and complicated. Relevant legislation is set out, respectively, (for England and Wales) in the Prescription Act 1832 and the Rights of Light Act 1959, and (for Northern Ireland) in the Prescription Act 1832 and the Rights of Light Act (Northern Ireland) 1961. A period of 20 years of uninterrupted access and enjoyment may be necessary to establish such a right. There is no equivalent legislation in Scotland.

It is possible to establish definite conditions under which the right would be upheld. A minimum sky factor of 0.2 per cent over one-half of the room at a working plane height of 850 mm is normally considered sufficient. (Note that the sky factor excludes reflected light.)

A general right to sunlight remains to be established in English law. There is no prescriptive right to a view.

Building Regulations

Approved means of satisfying the Building Regulations 1985 concerning the conservation of fuel and power include specified minimum standards of thermal insulation of the fabric and maximum areas of single glazing in walls and roofs. In some cases this specified maximum area may be less than that found necessary for effective daylighting. Procedures are specified in the Building Regulations 1985 which allow these glazed areas to be increased, but when this is achieved with glazing of lower thermal transmittance, allowance may have to be made for reduced transmission in the visible spectrum. To prevent the spread of fire to neighbouring premises the fenestration of external walling may have to be restricted in area depending on the purpose group, size and materials of the building, and its proximity to surrounding buildings or to the site boundary. The Building Regulations 1985 also include requirements for the ventilation of rooms.

RANGE OF TYPICAL LAMPS AND THEIR LUMEN OUTPUTS

Lamp type	Wattage	Average initial lumens (approx.)	Lamp type	Wattage	Average initial lumens (approx.)
Tungsten			*Low-pressure sodium*		
GLS	60	710	SOX	35	4 800
	150	2 180		180	3 300
	500	8 300	SOX-E	36	6 000
Halogen K	750	16 900		131	27 000
	1500	36 300			
			High-pressure sodium		
Fluorescent tubes			SON PLUS	150	16 000
White	70	5 740		250	28 500
	58	4 600		400	54 000
	36	2 850	SON deluxe	70	5 800
Krypton-filled 26 mm	100	9 400		150	12 250
Triphosphor	70	6 350		400	38 000
	58	5 200			
	36	3 350	*High-pressure mercury*		
			MBF	50	2 000
Compact lamps (PL)	15	900		125	6 500
	23	1 500		400	22 000
			MBIF	250	19 000
				1000	81 000

Note: Other lamp wattages/lumens can be derived from these data although high-wattage lamps normally have about 10 per cent higher luminous efficacy than lower wattages of the same lamp type, e.g. MBIF.

BIBLIOGRAPHY

A comparatively full list of references is contained in the CIBSE 1994 Code. The following are relevant books, CIBSE publications, recommendations and regulations that may be of help to the student. It is in no way a complete list and more information can be obtained from the current CIBSE publication *OPUS*.

Books

Lamps and Lighting, 4th edn, J. R. Coaton and A. M. Marsden (eds). London: Arnold 1997

Sight, Light and Work, H. C. Weston. London: Lewis 1962

Day Lighting, R. G. Hopkinson, P. Petherbridge and J. Longmore. London: Heinemann 1963

Lighting Fittings – Design and Performance, A. R. Bean and R. H. Simons. Oxford: Pergamon 1968 (to be revised)

Home Lighting, J. B. Harris. Newton Abbot: David & Charles 1984

Lighting, E. Effron. London: Guild Publishing 1986

British Standards and Codes of Practice

These will in time be replaced by ISO and CEN standards.

BS CP3   Chapter 1: Lighting. Part 1: Day lighting

BS CP3   Chapter 1: Lighting. Part 2: Artificial lighting

BS CP3   Chapter 1(B): Sunlight (houses, flats and schools)

BS CP3   Chapter 7(F): Provision of artificial light (houses, flats and schools)

BS 889   *Flameproof electric lighting fittings*

BS 1853   Parts 1 and 2: *Tubular fluorescent lamps for general lighting service*

BS 4533   Parts 1, 2, 101, 102 and 103: *Luminaires*

BS 4683   Parts 1, 2 and 4: *Electrical apparatus for explosive atmospheres*

BS 4727   Part 4: *Glossary of terms particular to lighting and colour*

BS 4800   *Paint colours for building purposes*

BS 4901   *Plastic colours for building purposes*

BS 5225   Parts 1 and 3: *Photometric measurements for luminaires*

BS 5252   *Framework for colour coordination for building purposes*

BS 5266   Parts 1 and 3: *Emergency lighting*

BS 5345   Parts 1–8: *Code of practice for the selection, installation and maintenance of electrical apparatus for use in potentially explosive atmospheres (other than mining applications or explosive processing and manufacture)*
BS 5489   Parts 1–10: *Code of practice for road lighting*
BS 5490   *Classification of degrees of protection provided by enclosures*
BS 5501   Parts 1–9: *Electrical apparatus for potentially explosive atmospheres*
BS 7179   *Ergonomics of design and use of VDTs in offices*

CIBSE publications

Codes for Interior Lighting, 1994 (commonly known as CIBSE 1994 Lighting Code)

Technical memoranda

TM5   *The calculation and use of utilisation factors*
TM10   *The calculation of glare indices*
TM12   *Emergency lighting*

Lighting guides

LG01   *The industrial environment* (1989)
LG02   *Hospitals and health care buildings* (1989)
LG03   *The visual environment for display screen use* (1996)
LG04   *Sports* (1990)
LG05   *Lecture, teaching and conference rooms* (1991)
LG06   *The outdoor environment* (1992)
LG07   *Office lighting* (1993)
LG08   *Museums and art galleries* (1994)
LG09   *Lighting for communal and residential buildings* (1997)
LGSSR   *Shipbuilding and ship repair* (1979)
LGHHE   *Lighting in hostile and hazardous environments* (1983)

Application manuals

AM02   *Window design* (1987)

Lighting Industry Federation Factfinders

No. 3   *Lamp guide*
No. 5   *The benefits of certification*
No. 6   *Hazardous area lighting*

ICEL GUIDE 1006   *Emergency Lighting Design Guide* 1997

Recommendations and reports of the Commission Internationale de l'Eclairage (CIE) (selection)

No. 13.2   *Method of measuring and specifying colour rendering of light sources* 1988
No. 17   *International lighting vocabulary* 1987

No. 18   *Physical photometry* 1986
No. 19/2   *An analytic model for describing the influence of lighting parameters on visual
              performance* 1981
No. 29   *Guide on interior lighting* 1986
No. 40   *Calculations for interior lighting – basic method* 1978
No. 46   *A review of publications on properties and reflection values of material reflection
              standard* 1979
No. 55   *Discomfort glare* 1984
No. 66   *Road surfaces and lighting* 1984
No. 97   *Maintenance*

BRE publications

BRER 288   *Designing buildings for daylight* (1995)
BRELP   *Designing for natural and artificial lighting* (1995)

RIBA/Thorn/CIBSE

RIBA01   *Electric lighting for buildings* (1996)

College of Optometrists publications

Visual efficiency in industry, J. W. Grundy (1980)
Light and eyes at work, P. T. Stone (1981)
Work and the eye, R. North (1993)

ANSWERS TO SELF-ASSESSMENT TASKS

1.1 508×10^{12} Hz

1.2 119 lm/W

1.3 ■ disability and discomfort
    ■ veiling or disability
    ■ mainly discomfort

2.1 intensity involves direction; the luminous flux does not define the intensity

2.2 220 cd

2.3 intensity and illuminance
    lumens fall on a larger area

2.4 inverse square gives 707 lx error + 57 per cent

2.5 recessed flat diffusing panel
    sphere
    vertical cylinder

3.1 (a) 5, (b) 3, (c) 4, (d) 2

3.2 grey; no

3.3 18 C 35; (x, y) 0.2, 0.2 (both approx.)

3.4 high-pressure sodium, warm white fluorescent

3.5 red/orange, sodium yellow, white/white

4.1 electron; no, in UV

4.2 reduce evaporation rate
    approx. double life
    to cope with 250°C temperature

4.3 see page 81

4.4 low-pressure lamps; create cool spot (fluorescent), thermally insulate (sodium)

4.5 does not produce significant amount of UV

4.6 outside light, e.g. bulkhead; more compact and not sensitive to cold temperatures

4.7 tungsten halogen

5.1 (a) diffuse, (b) mixed, (c) specular

5.2 (a) 2.0, (b) 6.8, (c) 2.4, (d) 4.4

5.3 intensity distributions, spacing to height ratios, LORs and DLORs, UF table

6.1 need at least 25 klx, approx. 1000 h

6.2 $(0.85 \times 8.2 \times 80)/144(1 - 0.18) = 4.7$ per cent

7.1 G will increase because $L_s^{1.6}$ increases by 1.6, but $\omega^{0.8}$ decreases by 0.78, so overall increase factor of 1.25

7.2   excessively bright ceiling
relatively inefficient
colour could be a problem
not instant restart

7.3   (a) no, (b) no, (c) yes

9.1   50 W/m^2, which is well above target load

9.2   the same as in Fig. 9.2

9.3   2 to 4 per cent

9.4   saving on cooling load
reduce heating requirements
cool the lamps
coordinate ceiling services

10.1   patches will narrow and become streaky

10.2   44.4 m

10.3   spacing becomes 12 m
maintenance may be easier, more columns needed

11.1   a good question! 15 per cent upward waste light permissible and luminance not to exceed 10 cd/m^2; luminance will conform if wall reflectances are below 20 per cent, waste light depends on how well the beams are trained

11.2   no – if you change the height you will need a different lantern to achieve the same illuminance and diversity

NUMERICAL SOLUTIONS
TO EXERCISES

1.3   B 36 per cent higher efficacy

2.1   external luminance 365 cd/m^2
       illuminance 8.0 lx

2.2   117 lx

2.3   944 lx and 66 lx

2.4   1820 and 2777 cd/m^2

3.3 x is 0.265, y is 0.270

3.4   double

4.2   1.14 H

4.3   (a) no significant difference
       (b) PL lamps will be more economical after 5 years

5.3   22.2

6.1   1.66 per cent

6.2   11 per cent

7.3   Approx. 28

9.1   (c) target load (from Table 9.1 and room index of 3.3) for 300 lx is 3.7 W/m^2 based
          on UF 0.46 and MF 0.7
       (d) circuit watts approx. 2.4 kW; if 70 per cent is recoverable, 1.68 kW
          Note: both these sets of calculations are approx.

10.2   3.6 lx

INDEX

acuity 8
air handling
   luminaires 207
   performance 208
apostilb 44
apparent area 22
aspect factors 41
average daylight factor 140

beam diagrams 35
black body radiator laws 65
blinking 13
British Zonal (BZ) Classification 26
building regulations 252, 253
building types
   factories 188
   home 193
   offices 189
   shops 192

calculux 246
candela 18
CIE chromaticity diagram 54
CIE luminance curves 172
CIE overcast sky 125
classification of luminaires
   BS 4533 115
   BZ 26
   CIE 28
   cylindrical illuminance 160
   electrical protection 115
   flammable atmospheres 114
   Ingress Protection (IP) 115
   photometric performance 118
   supporting surfaces 114
colour
   additive mixing 50
   appearance 50
   BS 5252 52
   CIE chromaticity diagram 54
   constancy 60
   correlated colour temperature 53
   matching 60

metamerism 60
Munsell Colour Atlas 52
perception 10
primaries 51
rendering index (R_a) 59
subtractive mixing 51
uniform chromaticity scale (UCS) 57
compact flourescent lamps 85
computer programs 245
cosine effect 32
cylindrical illuminance 160

daylight
   average daylight factor 140
   CIE overcast sky 125
   factor, roof rights 143
   factor, side windows 127
   losses 136
   no–sky line 133
   orientation factor 141
   sunlight 145
   uniform sky 124
   variation 126
   window design 142
dichroic coatings 113
diffusion 112
dimming of fluorescent tubes 80
disability glare 13, 166
disc source illuminance 36
discomfort glare 13, 166
DLOR 30

EEC Ditectives 251
electromagnetic
   radiation 2
   spectrum 4
emergency lighting
   layouts 182
   maintained 183
   non-maintained 184
   signs 184
energy management
   compact fluorescent lamps 85
   energy saver lamps 209

heat recovery 204
integration with daylight 203
planned maintenance 200
'plug-in' lamps 209
switch control 202
target-lighting loads 198
uplighters 200
eye
cross-section 5
performance *see* vision

flat diffuser 45
floodlighting
beam angles 236
illuminance levels 239
lumen calculation 239
photometric data 238
planning 236
uniformity 236
fluorescence 73
fluorescent tubular lamps 76

glare
calculations 166
CIE method 171
disability 13
discomfort 13
index 167
large areas 171
LG3 categories 190
VDS areas 190
veiling 13

heat
distribution from lamps 76, 204
irradiance 204
luminaire distribution 206
recovery 204
high-frequency operation 80
high-pressure lamps
mercury 88
metal halide 91
sodium 91
HYPERLITE 3 249

illumination
aspect factors 42
beam diagrams 35
cosine effect 32
cylindrical 160
daylight variation 126
direct 31, 161
disc source 36
flow chart 150
inter-reflected 180
linear source 39
point source 30
rectangular source 44
room surfaces 162
scalar 159
standard maintained values 149

Ingress Protection (IP) 115
intensity distributions 24, 152
inter-reflection 180
inverse square law 31
isocandela diagram 26
isolux diagram 33, 179, 229, 241

lamps
filament (GLS) 66
fluorescent, compact 85
fluorescent, tubular 76
high-frequency operation 80
induction 85
mercury, high-pressure 88
metal halide 91
sodium, high-pressure 91
sodium, low-pressure 85
summary tables 96, 255
tungsten halogen 68
layouts
general 155, 157
local 158
localised 158
spacing 154
Legal requirements
Building Regulations 252, 253
EEC directives 251
rights of light 252
UK legislation 251
types of energy saving 198
LG3 190
light output ratios 30
Light pollution 231
light production
discharge through a gas 71
fluorescence 73
halogen cycle 68
radiation laws 64
linear source illuminance 39
local lighting 158
localised lighting 158
Louvres 108
lumen 18
Lumen maintenance 97
luminaires
dichroic coatings 113
diffusers 112
electrical protection 115
Fibre optics 113
uplighters 146, 170
finishes 101
flammable situations 114
Ingress Protection (IP) 115
materials 100
photometric performance 118
reflectors 101
refractors 113
uprighters 146, 170
luminance
definition 19, 21
diffusing surface 45

distribution in rooms 162
glare, CIE curves 171
photometric data 152
roads 217
sky 102
luminous
efficacy of lamps 96, 202
intensity 18, 19, 20
flux 18, 23
lux 18

maintained emergency lighting 183
maintenance factor
daylight 138
lamp lumen 153
lamp survival 98, 153
luminaire 154
roadway 222
room surface 154
materials and finishes 100
mercury lamps 88
metal halide lamps 91
Munsell Colour Atlas 52

no-sky line 133
non-maintained emergency lighting 184

photometric
data 151, 220, 236
performance 118
photons 72
Planck's equation 64
planned maintenance 200
'plug-in' lamps 209
point source illuminance 30
polar curves 24
presbyopia 10
primaries (colours) 51
protractors (BRE) 128

radiation laws 65
rectangular source illuminance 44
reflection laws 102
reflectors 101
refractors 109
relative eye sensitivity (V_λ) 6
rights of light 252
roadway lighting
bends 223
BS 5489 214
glare 213, 216
junctions 223
layouts 218
light pollution 231
luminance 217
maintenance factors 222
photometric data 220
road surfaces 219
silhouette vision 212
subsidiary roads 225
room index 151

scalar illuminance 159
secondary colours 51
semi-resonant start control circuits 79
sky
CIE overcast 125
component 127
luminance 124
uniform 124
sodium lamps
high pressure 91
low-pressure 85
spacing/height ratio 154
spectrum 4
starterless control circuits 78
subtractive colour mixing 50
sunlight 145
survival of lamps 98, 153
switch control of lighting circuits 202
switch start control circuits 77

target lighting loads 198
temperature
correlated colour 54
effect on fluorescent tubes 82
radiators 64
UK legislation 251
transfer functions 176
transmittance of window glasses 137
tubular fluorescent lamps 76
tungsten filament lamps (GLS) 66
tungsten halogen cycle 68

ULOR 30
uniform sky 124
uniformity 154, 215, 236
uplighters 174, 200
uplighting
equivalent point source 176
inter-reflection 180
lumen method 175
utilisation factors 176
utilisation factors 151, 176, 229

vision
acuity 8
colour 10
contrast sensitivity 10
glare 12
optics 8
performance 8
preference 15
presbyopia 10
relative response 6
task requirements 14
visualisation 246

wavelength 4
window design 142

zone factors 29